South Field

Copse

West Field

Barn

North Field

North

*For Juliette, Daphné, Alexandre
and Célestine.
You are my true north.*

ANGELS *in* *the* CELLAR

Peter Hahn

LITTLE TOLLER

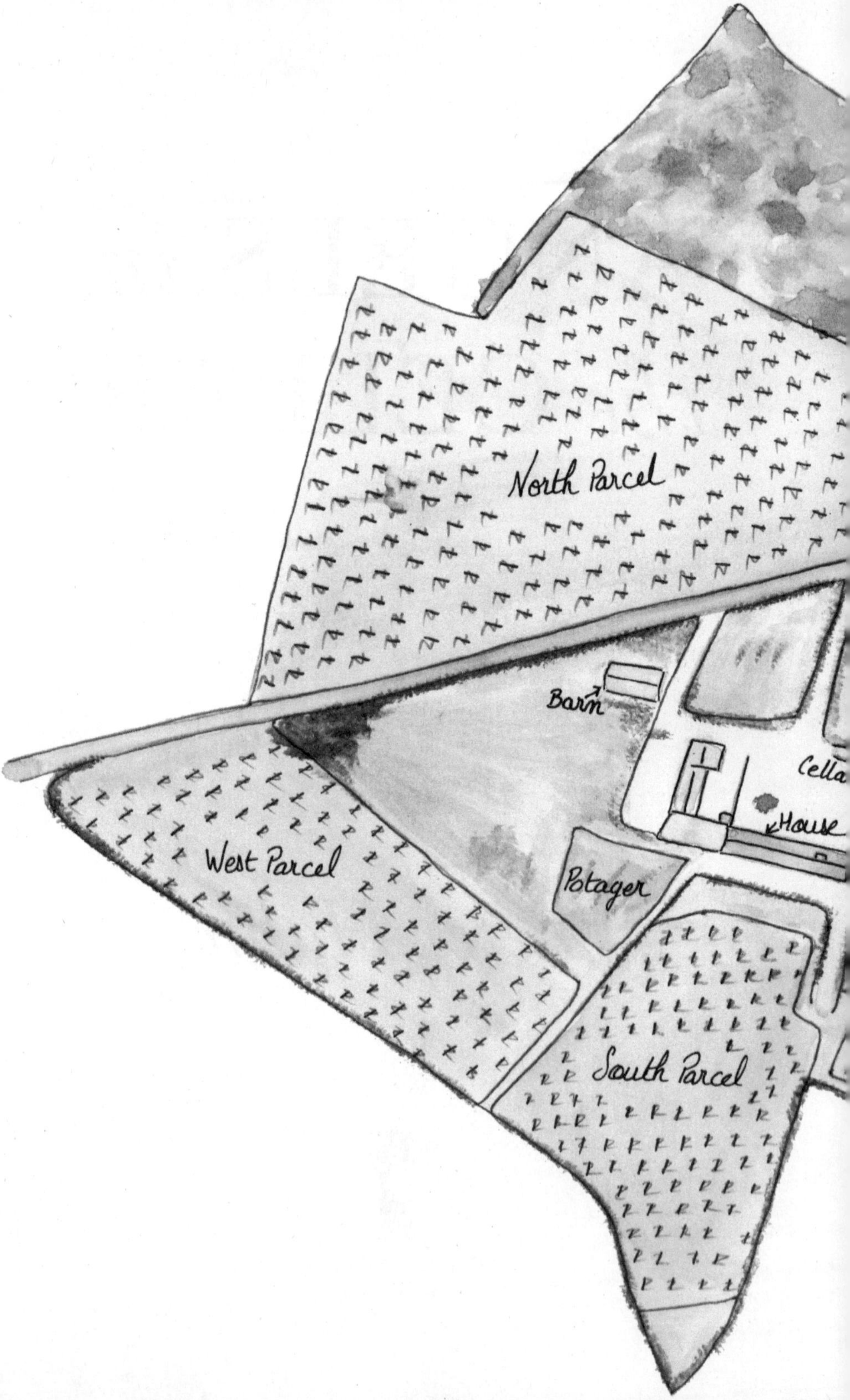
North Parcel
Barn
Cella
House
West Parcel
Potager
South Parcel

Le Clos de la Mesterie

Woods

Winery

Sequoia

Pine

East Parcel

North

Prologue

When we try to pick out anything by itself,
we find it hitched to everything else in the universe.

JOHN MUIR

For the last two decades I've lived and worked on a small vineyard in the Loire Valley, called Le Clos de la Meslerie. I arrived here in 2002, and have rarely left since. A four-hundred-year-old pile of a house, crumbling, untouched for decades, encircled by ten acres of woodland, fields and – crucially – 'Appellation Contrôlée Vouvray' vineyard, also very much in need of renovation. After twenty-three years of caring for this place, nursing the land and buildings back to health, I realise it will take twenty more, and then twenty again.

It was sometime near the turn of this century when my world collapsed around me. I was in my late thirties, sitting in a black cab on my way to Heathrow Airport, laptop open on my knees, Blackberry in one hand and mobile phone in the other. It was Thursday afternoon. The sky was an evanescent blue. I had just worked for 48 hours without sleep, and was heading home to recover before my next meeting the following morning. The cab drove into a tunnel and the window was transformed into a mirror; I stared at the man reflected but didn't recognise him. It was only for a second or two, but it was enough to make me feel vaguely sick. My fists were in tight balls, my body rigid. I felt a floating sensation. The cabbie was asking me something, the words were barely comprehensible, *You a'right mate?* I turned to see his face in the rear-view mirror but couldn't speak. I was far from being alright, and would remain that way for some time. There was something I was longing for, but didn't know where it was or how I could ever get there.

There is a word in the Celtic languages that I love, for both its sound and its meaning. In Welsh it is *hiraeth*, in Breton *hiraezh*. It describes a feeling of nostalgia or yearning to go back to a home or a place that you can't return to, because either it no longer exists or it never existed in the first place. It is both the real and imagined bond we feel with a landscape, a time, an era or a person. This is how I felt. I knew there was a place I was longing for. But how was I suposed to find it if I could no longer return there or it never even existed?

As I mulled over these thoughts, it started to become clearer that any sort of meaningful life for me was somehow linked to making a living from the land. I didn't know why or understand how. It was like flying completely blind into the sun, with a low-humming instinct. I knew things had to change, and that for me this change would be quite drastic.

Gradually, the notion of trading Power-Point slides and Excel spreadsheets for something more elemental became intoxicating. And so much the better if what I did with the land also made other people feel *good* about themselves or touched something basic about being human. It could have meant becoming a baker, carpenter or candlestick-maker. But my choice was to become a farmer. Or, more precisely, a vine-grower. I had been a wine-lover and general epicurean for years, and for a time I played with the idea of opening a wine shop or becoming a wine importer and distributor. But now, deep down, for reasons beyond me, I knew that the land was calling.

From Paris, I set out into the winelands of France. For almost two years I spent all of my spare time roaming vineyards in the regions I was familiar with: the Rhone, Languedoc, Bordeaux,

Provence. I found Touraine, in the Loire Valley, completely by chance. The more I read about it and tasted wines from the region, the more intrigued I became and started to feel something like a familiarity. Touraine's climate was the most temperate of all the regions, with mild winters and summers, but with distinct seasons nonetheless. It was green and lush from regular rainfall, with gently rolling hills and valleys carved by the Loire and other tributaries. It is known as the 'Jardin de la France' because its fine soils and gentle climate make it ideal for growing a broad basket of vegetables and fruits, including grapes, almost all year round.

I decided I had to see it, and when the train from Paris crossed the broad expanse of the Loire River, a sense of ease grew in me. I felt welcome. An estate agent picked me up at the station and we drove twenty minutes to the farm. We didn't speak much along the way; I wanted to look, to really take in my surroundings: the undulating terrain, the small roads winding off towards little hillside hamlets. 'We're almost there,' she said. 'It's just up this hill.'

When we arrived, it felt like Le Clos de la Meslerie had been waiting for me. More profound than love at first sight and somehow very practical. The feeling was simply: *You and I can work together.* Rationally, I knew it was just stone, slate tiles, old windows with oak frames, paint peeling off everywhere, ivy growing up into the gutters and onto the roof. I knew it was just grass and trees, gravel, overgrown fields and a dilapidated vineyard. I knew all of this. But I knew something else, too. Finally, I had found *it.* A place where I could become myself.

There are several names for what I do now. The most common is probably 'winemaker'. In the beginning, that's what I called it, too. But with time and experience I now use the word *winegrower.* I know now that wine is, to a much greater extent, grown rather than made. This distinction between growing and making is at the heart of a sea-change that has been taking place in the world of wine for decades now: wine is made outside in the vineyard and whatever happens in the cellar is ancillary. Everything is in the grapes. I am, first and foremost, a grape farmer.

My first vintage was in 2008. The previous winegrower had died decades before and his wife was unable to manage the land on her own, so the family had rented the vines to a local grower who would grow the grapes and take them back to his winery to be mixed in with his harvest. When I moved into the house in January 2002, there were still several years remaining on the lease, so I went back to school to do a degree in viticulture and oenology, getting myself ready. I read voraciously, technical books and papers, but also literary and philosophical works, stories of winegrowers, histories of wine and, perhaps most importantly, I spent time with neighbouring winegrowers, listening. I worked with them. I watched them. I asked questions. All of this was essential in honing my approach to growing.

Along the way, one of the greatest personal discoveries I've made since becoming a winegrower is what it means to live *with* and *in* the seasons. In my previous life, I would jump in a plane to *escape* from autumn and winter – the dark times, the depressing times. But they are depressing only when you fight them. They become heavy and gloomy when you are sitting in an office somewhere – you arrive there in the morning darkness, and you leave after night has fallen. The cold, the rain, the snow are disruptions, an assault on anyone in an Italian suit and shiny leather shoes, completely unfit to face these elements. When you work on the land, however, the seasons are not separate from you. They are part of what you do. They guide your activities, remind you of what needs to be done. The seasons dictate the right clothes to wear and are absolutely essential to being at home in the place where you work.

The seasons can be capricious, too. They can be like the closest of friends – true confidants, brothers in arms. Then the next day, these friends rage against you, uncooperative, rude, flippant, downright annoying. But as you develop your relationship with the seasons, you listen, you watch, you learn to relax, and instead of struggling against nature, pushing back and trying to manipulate how plants grow, exerting control, you soon realise that there are many, many more good turns than bad.

L'HIVER
WINTER

Plants and animals don't fight the winter; they don't pretend it's not happening and attempt to carry on living the same lives that they lived in the summer. They prepare. They adapt. They perform extraordinary acts of metamorphosis to get them through.

KATHERINE MAY, *Wintering*

Since I have been living on the vineyard, I have found that most seasons pass far too quickly. I try to hold onto time by its mercurial strands, pulling it back and striving to stretch the days out; but they almost always contract. Winter is the only season that spreads itself out, sending out a call to slow down, to rest. The short days provide long nights, which are an excuse to read, sleep, contemplate.

But winter does not mean there is free time on a vineyard. In fact, winter is the beginning of the year, not the end. It's the time when a new vintage is conceived, and therefore everything I do on the vineyard in those cold months is preparation for the next harvest. With the end of autumn, my mind turns to endurance, my body to marathon mode. I know the winter will be hard. Yet the paradox of winter is that, though it can be physically draining and mentally challenging, it also bestows plenty of recovery time if you handle it with care.

The main winter activity on a vineyard is the pruning and shaping of the vines. Pruning is deliberate, repetitive and unhurried. I need to pace myself because on every acre of this vineyard there are over 2,500 vines, making for a total of close to 24,000 individual plants. And I will need to take care of each and every one of them, looking closely, taking stock before a single cut is made, before I carve each one for the next growing season. Pruning is all about making decisions. Taking into account a plant's strength and vigour, its age, even its position in the vineyard, I shape the vine according to how many bunches of grapes I think it should grow in the months to come. It's the moment when I make a choice about the quantity and quality of the grapes I am asking the vine to give me. And I resist hiring people to do this job with me, because it's perhaps the most important and rewarding activity on a vineyard. Every year, without fail, pruning teaches me to live in slow motion.

When you look at a vineyard from a distance there is something mesmerising about it: row upon equally spaced row of vines fading into the distance, like the Terracotta Army of Qin Shi Huang, the first emperor of China, thousands of soldiers in perfect formation. But this big picture is misleading. As you get closer you see that each vine, like each terracotta figure, is completely different; a different face, a different position of the arms or hands, each character wearing time in a different way. So it is with vines in a vineyard. They are similar but not the same. This is especially true in an old-plant vineyard. At Clos de la Meslerie, our vines average around 60 years of age, with many pushing 80 or 90 years. Modern, intensive wine production usually requires vines to be grubbed up and replanted after 30 or so years, when their productivity begins to wane. Often, vineyards will be grubbed up and replanted with a different grape variety, responding to consumer trends. If Sauvignon is popular, let's plant it.

In the established vineyards of France, however, this is not the case. Old vines are respected and preserved and not treated merely as fruit farms as they often are by many New World producers. Here, they are history. They are cultural treasures, the ongoing story of family, farm and village, of the whole region. Those thousands upon thousands of vines marching across the hills and valleys of French wine-growing regions are part of the local population. I like to think that that is where the winegrowers themselves go, and where I will go eventually, reincarnated as one of those vine-soldiers.

It was in the middle of January when we arrived on the farm. There have been many Januarys now but, for the first few, this cold, old house was both sanctuary and burden. On many a January night, trying to keep warm near a fire in one of the draughty rooms, after a day of stripping wallpaper or painting ceilings, and often discovering unpleasant surprises wrought on the house by time, my wife, Juliette, and I would pull out what we called 'the old box'. The box has now been emptied, but for many years it was a symbol of all those who had come before us in this place.

A few days before the final signing of the purchase contract with the Sauger family, the matriarch, Suzanne, invited us to the house for an early evening *apéritif*. We sipped wine from the vineyard, a fire crackling in the hearth of the sitting room, somewhat smoky as the draw seemed to be struggling. I looked around at the old family photographs on the walls and side tables. Candles had been lit throughout the room because the electricity had shorted out that morning. The furniture was elegant but worn, the wallpaper ornate and faded. It could just as easily have been January 1902 as January 2002; it would have looked and felt the same. As I watched flakes of snow scurrying across the window panes, I realised that I was not just moving house. I was coming to a *place*, and that these people in this room with me were as much a part of that place as I ever would be.

After we'd drained our glasses Suzanne touched my arm lightly and asked if I could lift a box down from the shelf she was pointing at. I obliged and put the old box on the coffee table. She told me to open it. Inside, it was filled with what looked to be simply a stack of old paper. Suzanne explained that every owner of the property, for centuries, had passed the papers in this box on to the new owner. They were all the legal documents linked to the place: deeds, ownership rights, land transactions, debt agreements and sales titles from whenever the farm changed hands. At that moment, while I was touched (and told Suzanne and her family as much), I didn't quite realise what the old box meant. It was only over the following years, as Juliette and I took the time to look through the documents one by one, that I began to appreciate its significance, and I will certainly, if the time ever comes in my lifetime, do the same honour to the next person and family who takes up stewardship of this farm.

All the papers in the box were a written emblem of the deep and binding link between this place of stone, mortar and soil and the community of people who worked here, along with their animals, trees and vines. These original, notarised documents, most written with a quill, later ones with a fountain pen and then typewriter, embraced the history of this place. The first document was dated 1621, the odd flourishes of the quill making the ornate writing illegible to our modern eyes. The more recent documents were easier to read and revealed the families who were here before us: the de la Richières, Larribe-Bourdas', the Archambauds, the de Forestières. From this old box, our predecessors became human.

Students of French history will know that after the revolution of 1789, the zeal for creating a new society and razing the past included the rejection of the Gregorian calendar and a creation of a new one – the French Republican calendar – which eliminated all religious and royalist influence. It marked the start of a brave new world, and on 21 September 1792, Year 1 was declared. On the calendar, the names of the months were changed to reflect the land and the seasons – thus, the equivalent of May became Floréal, derived from the Latin, 'Flower'; August became Thermidor, to indicate 'heat' and so on. Each month was divided into three weeks of ten days, with each of these days simply numbered – the fourth day being *quartidi*, for example. All official documents were dated using this system for thirteen years, until the revolution expired in a blaze of empire, and Napoleon was crowned in 1804.

On a February night, after we'd moved in, Juliette and I came across an odd document while we were sorting through the old box, arranging the papers by century, then year. On it, a certain 'Jean Johan' residing at 'La Meslerie' agreed to acquire a parcel of land nearby in the interest of growing grapes. It was signed and dated the 27th day in the month of Prairial (Meadow), Year 12. The fire crackled and spat against the worn stone fireplace, suddenly ejecting a small glowing ember out into the room, causing Juliette and I to recoil as we slipped the document into a cardboard sleeve. Below the revolutionary date on the sleeve was written, *Vineyard plot acquisition, 15 June, 1804*. And for that moment we could imagine Monsieur Johan, winegrower, sitting there with us, before this fireplace, where he had surely sat with his family in the Year 12.

Winter always gets me thinking in long patterns. The slow days of pruning, more than any other time of year, give me time and space to contemplate what I'm doing and why I'm doing it. Not only that, but also *how* I am doing it. I learned an uncomfortable truth in my early years in the wine world: all wine is *not* created equal. Approaches to growing grapes and making wine are as diverse as culture, scattered across an intricate, multi-dimensional map.

It is fair to say, however, that there is also a simplified, two-dimensional map: at one end of the scale there is 'industrial' growing and at the other end 'natural'. The gulf is enormous. It is equivalent to the difference between a machine-produced polyester sweater and one hand-knitted in Aran or Cumbria, made of locally produced and spun wool, with the name of the knitter on the label. The hand-knitted sweater will not be machine-perfect; it will be irregular, because each hand-knitted sweater is a unique creation, linked to the person who made it and the materials that went into it. While industrial sweaters will always be the same, mass-produced and regular, sold in stores all over the world, cheaper for sure, the chances are that the hand-made sweater will have patterns that you will find nowhere else. It will also likely outlast the machine-made sweater by decades.

Our wine is hand-made. It is the unique product of a place and of the people who nurture the vines and the wine. It is made with as little machine and chemical intervention as possible. Nature is the driving force behind it. This is not to say that I believe that it is 'better' than other wines. But, like the hand-knitted sweater, it is unique. It is an expression of the place. I know that we can always do better, make more effort not to intervene or to intervene more judiciously – every vintage causes this soul-searching, and there is always a feeling of discontent with the result. Fortunately, each year the discontent is always outshone by the joy of the new vintage.

In winter, when I contemplate what I am doing and why I am doing it, I always come back to the reason I made this journey from the back of a cab in London to the Loire Valley, and the visceral need I have discovered to feel a connection to this land, to its life and cycles, and to not only understand my place in it but also to participate actively in it. To be an actor rather than a spectator. The human-made world of cities and commerce, of reason and science, of philosophy and art, are all also fascinating to me, and I do not reject it. But I am much more critical of it now, and very selective in what I allow in or go out to seek. And so, my activity on the vineyard is guided by the principle that, in general, nature will tell me what to do if I listen carefully enough.

In my work here, I maintain a healthy scepticism about new technologies and approaches constantly peddled to winegrowers. I remind myself that people have been growing grapes and making wine for many thousands of years – the latest evidence shows wine being made 8,000 years ago in what is now Georgia, on the Black Sea. Today's intensive, industrial farming is a twentieth-century invention. Herbicides, pesticides, synthetic fertilisers: all are recent 'tools'. Wine 'technology', therefore, has pretty much only been with us since the 1950s, and the science of oenology has only really taken hold of winemaking in the last 50 years.

So how did we do it for those thousands of years before? And what did wine taste like 3,000, 1,000, 500, or 100 years ago? This is not to say that I believe that everything was better in the past, or that I am a romantic nostalgic with no grip on the practicalities of the world. It is simply a question I bear vaguely in the back of my mind as I go about my business every day.

There is never a pruning season when I don't think about the two people who first taught me about the physiology of vines: Vincent and Damien. I met them both on the same day almost a quarter of a century ago. I was a youngish man then, still in my thirties, but they were very young men. Barely into their twenties. When I met them I did not know that over the coming decades I would attend their thirtieth and then fortieth birthday parties, dance at their weddings, be there to celebrate the births of their children, be the godfather of one of them, and attend the funerals of their parents. I would see them grow from young men setting out on their life's paths to becoming two of the most reputed and respected winegrowers in the region, owning and running two of the most prestigious wineries in the Loire Valley, and producing award-winning wines sold in the four corners of the world. I didn't know that they would become my teachers, my mentors and my dear friends.

I arrived for my first day of school with butterflies in my stomach, hardly able to pronounce the mouthful of the school's name: *le Lycée Professionnel Agricole d'Amboise*. It was one of many such institutions scattered across France, a key link in the chain of the continuity of agriculture. Being an apprentice or skilled worker is not frowned upon here, at least not outside of the major cities. The countryside is still proud to send their children to be educated in stewarding the land. And most of these colleges and schools offer continuing education degrees for graduates of the traditional lycées. It was in one of these programmes that I was enrolled, setting out to get my *Brevet Professionnel Agricole* degree. A two-year full-time programme. The descriptive brochure promised me, upon completion, that I would have acquired the competencies required to be a *Responsable d'une exploitation viti/vinicole*. I would be a qualified winegrower, entitled to own and manage a wine estate. The very thought of it seemed impossible to me, unreal.

That first day is mostly a blur to me now, but I remember two classes, and meeting the teachers, as if it were yesterday:

Viticulture: Soil Constitution – Vincent Carême
Viticulture: Grape Varietals – Damien Delecheneau

In many ways Vincent and Daminen were opposites. Vincent had dark hair and dark eyes, and was reserved to the point of seeming cold. His hair was cut short, almost military in appearance. He seemed to me at that moment to be something of a caricature of a French farmer. One could easily picture him in the classic blue work overalls, wearing a beret and a five o'clock shadow, a smoking Gaulloise hanging from the side of his mouth. Damien was fair and blue-eyed, had long, tousled hair and looked like he'd just got out of a convertible. His countenance and easy manner made me think of a California beach boy. I would learn, over time, that their differences were skin-deep and their similarities soul-deep. Their kindness, generosity, sense of rigour, work ethic and deep love of the land and its produce inspired me, guided me and became one of the threads that wove my project and my life on the land together.

The long hours we spent together, first in the classroom and later in their vineyards, cellar, and then finally, my vineyard and cellar, are now imprinted on both me and this place.

Damien taught the pruning module at school. I remember being out in the school's educational vineyard with him and my fellow students in January. Watching him prune was, for me, like watching some strange and exotic dancer, reminding me, in a remote way, of Kathak dancing of northern India with his speed and precision, bending his body and twisting his arm and wrist. I don't think I have ever attained the skill to prune as quickly and accurately as he did on that day, but I have surely tried, and have made progress over the years nonetheless.

Pruning the vines is in many ways 'the final frontier' of traditional winegrowing. It is perhaps the last of the many activities in the quality French vineyards that will be performed by machines. Because every vine is different and the cuts you make are specific to every single cane on every single vine, the uniformity required for machines is still a stretch for technology. For now. There are of course tractors and tools for ploughing, machines for spraying, pre-pruning, even de-budding, leaf thinning and for picking the grapes. As I write, GPS-controlled, self-driving vineyard tractors are being perfected and pruning robots are being developed. Who knows what AI will do. When this all happens, it will be final. People will rarely, if ever, have to venture into the vineyard again. I often wonder what that world will look like. It's not very far away, so I might even see it. It will turn me into one of the last dinosaurs.

But how will we replace that feeling? The richness and life-affirming energy of working on the land? Will there be virtual reality technology that will allow us to experience this without leaving our desks or even our beds?

Humankind's eternal quest to remove itself from the land, to separate itself from and dominate nature will then be complete. Perhaps I am too pessimistic, but what I have seen, working as a traditional organic farmer for the last sixteen years, is not promising. It is true that there has been a movement towards organic, locally sourced, low-carbon agricultural production and consumption, but this is anecdotal, a 'lifestyle' choice, a small pothole on the road to everything being mass-produced and managed by machines.

The human striving to get away from the land is of course tied directly to our desire to avoid work that is physically demanding and repetitive; work that wears against the body, grinds the joints and strains the muscles. Work that leaves a person spent at the end of the day, wondering if it will even be possible to start again the following morning. Yet for me, this is what brings me back to pruning.

The spirit of each day is intimately linked with the weather and mood. Because the pruning season stretches from December to March, the character of the task wears different faces. There is deep-winter pruning, there is late-winter pruning, and there is early-spring pruning. So this work, seemingly so monotonous, repetitive, and bathed in sameness, is actually more diverse than one might think. The one thing that doesn't change over the course of the season is that every single cane must be cut. This is the ineluctable truth, and it hurts.

There are now electric secateurs powered by a battery pack. And there are pre-pruning machines that go down the rows cutting a straight line through the top three-quarters of all the canes. This pre-pruning is done because, while cutting the cane is one thing, the harder part is pulling the long cane out of the trellising wires. The canes wind themselves into the trellising and hold on surprisingly tightly by the tendrils that weave around the wires. Pre-pruning lessens the problem somewhat, but it requires going through the vineyard on a tractor with a pre-pruning machine attached. I am extremely prudent about using my tractor. Every time you go through the vineyard with a tractor (even though I have the smallest and lightest tractor available) you compact the soil, smothering it to an extent. Also, you inevitably break some vines, crush anthills, and burn diesel fuel. The noise of the tractor and the clatter of the machine are an additional, if only aesthetic, downside.

So I do not pre-prune. I take the time and expend the energy necessary to do without. Nor do I use electric secateurs. The muscles of my right wrist and forearm sometimes scream out at me, *Go electric*, but I resist. For all the gains that technology can bring, my sense is that the subtle, sometimes almost imperceptible, losses that follow are not worth the gains. I want to emphasise again that this is *not* virtue signalling. This is just the way we think and work here.

Every vineyard has its own approach and, most importantly, its own constraints: people-power, economics, logistics, and so on. Every winegrower must find his or her way. But tolerance, on my part, stops at the extreme ends of industrial winegrowing, where the extensive use of chemical pesticides, herbicides and fertilisers is commonplace. This is unforgivable and unconscionable given what we now know about the damage it does to the environment and human health. Fortunately, this awareness is growing in French winegrowing circles and in the Loire Valley in particular, so the use of herbicides and pesticides is becoming much more judicious.

What could one possibly lose by choosing electric over manual secateurs? At its worst, pruning is drudgery. But, as I intimated earlier, at its best it approaches a type of art. At this level, the difference between electric and manual secateurs can be likened to using different-width paint brushes to paint a picture. There may be many cases when the wider brush suffices: you can cover the canvas more quickly, but some of the finesse and detail will be lost if you don't switch to a finer brush. Electric secateurs are larger and bulkier than manual ones, and therefore it becomes tricky making tight, difficult cuts. The supple human wrist, paired with manual secateurs, is a remarkably accurate instrument. It becomes less so with a bulkier tool. But even more than this, you lose feeling with electric secateurs. As you cut a cane with manual secateurs, you feel the cane through the tool. You feel how much resistance there is, how thick it is, and therefore how vigorous the vine itself is. You can use your eyes to see this, but to feel it is different. You can make better decisions about how to prune an individual vine. On a less vigorous vine, you might prune a little more aggressively in order to get the vine to produce fewer bunches next season. On a more vigorous vine it would be the opposite, and you might even prune to create a new arm for the following season, making the overall vine larger and allowing it to produce more grapes.

In deepest winter it doesn't get light in the Loire Valley until almost 9 a.m, and because you need light to work in the vines, there's time to ease into the morning. I usually sleep a bit later than in other seasons, but once up the first thing I do is bake bread if we need it. Then I might make a fire in the office, put some music on and, if I am feeling indulgent, sit in my armchair and read a novel for an hour. More often than not, though, I go to my desk, glance at the world news on my computer (a vestige of my previous life) and do administrative work.

French society is rather heavy on paperwork. Blame it on Napoleon. The French themselves often complain and joke about it, and the wine-world is no exception. *Au contraire*, because wine is alcohol and therefore subject to special regulations and taxes, the admin piles up. When I started, it involved a Kafkaesque shuffling of piles of paper. Fortunately, almost all of it is now online and at the mercy of my keyboard, and this is one technological advance that I relish because pretty much every move you make on a vineyard must be defined and declared, tracked by agencies, organisations and authorities. Every time you sell wine, you have to tell someone. Each harvest must be reported: the amount of crushed grapes and stems you have produced, and of course how much wine you have stored in the cellar. Is it in bottles or barrels or vats? If you grub up a parcel for replanting, how many vines will you replant? It goes on.

I have learned to accept all of this with some grace. After all, these are things I cannot change. I have also made an effort to understand why things are like this here. Of course, part of it is to make sure that the government does not lose out on alcohol-related tax revenues. But there is a deeper and somewhat more admirable purpose, too, which has something to do with the French notion of the state and its role in providing quite extensive protection for its people. It relates to structure and rules being set for things to function fairly, to avoid fraud but also, most importantly, to ensure a certain level of quality.

This is particularly true of the *appellation d'origine protégée* (AOP) system, which defines food and drink according to the geographic region it comes from and the traditional way in which it is made. Wine, cheese, meat, poultry, honey, fruit: all are categorised according to these local quality-control rules. In the case of wine, these rules get down to very specific details, such as restricting the number of buds you can leave on a vine when you're pruning, thus limiting the amount of wine you are allowed to produce from an acre of vines. The thinking behind this is simple: if you push vines to overproduce, there will be a decline in the quality of the wine, which has an impact on the identity and reputation of other vineyards in the region. Constraints like these are unheard of in the vineyards of the New World, where winegrowers see all this as slightly absurd, as do a growing number of French winemakers. These 'wine rebels' say it's the individual who should be responsible for the wine they make, not the wider community; this is an issue of long-standing discussion here. There are good arguments on both sides, and in most of these questions, like so many things, I believe the answer lies somewhere in the middle. However, I for one appreciate knowing that when I buy something stamped AOP it has been held to far stricter quality and environmental standards than most non-AOP produce or produce that comes from countries with virtually no quality and environmental safeguards.

At Le Clos de la Meslerie, the house and winemaking buildings, the cellar and barn, are all directly at the centre of the farm's 25 acres. The ten acres of vines that cluster uniformly around these buildings are essentially four pieces of land which we conveniently, if somewhat unimaginatively, call the North, South, East and West parcels. But the nomenclature of four compass points does actually have more significance than simply locating them relative to the house and other buildings, because the farm is on the top of a hill and the orientation of each parcel gives it a different exposure to the sunlight, wind and water.

My office is on the east side of the house, so in the morning, as the sun breaks through the winter darkness, I am well positioned to feel the first hints of a new day, and as the light infuses the East parcel, this tells me it's time to get ready to prune. Looking out of the large doors in the direction of the sunrise, I can see the bare canes of the vines reaching to the sky, turning a very light rose-coloured hue.

On the lawn between the house and those eastern vines, a giant sequoia also catches the first rays on its lush, light-green coat. While I take great pleasure in the dense woods on the farm, there are many individual trees dotted around that I love, and this sequoia is probably my favourite. Giant sequoias are the most massive living things on our planet. They can rise to 90 metres high and live for over 3,000 years. To me, they feel like the sentinels of the earth. The general majesty and dominance of these trees are undeniable, but there are other, more subtle and gentle characteristics too. The bark of the sequoia is a beautiful red brown; soft and fibrous, almost spongy and smooth. This makes it the perfect target for the tree-hugger in all of us. While the species does not shed its foliage seasonally, over the years wind and other weather cause many bits of the foliage to fall, building up into a springy mattress-mound of decomposition that extends out from the base of the tree. I have spent many an hour in the warmer months lying on this soft ground, reading or looking up at the massive branches that filter the sky. Our tree is over a hundred years old and stretches 25 metres high. It would be even taller if a lightning bolt hadn't struck it sometime in the mid-twentieth century, lopping off a third of its height. It is both damaged and beautiful.

Our sequoia is also special to me because we share an origin story – we are both originally American. There were sequoias in Europe until the last ice age, but all Giant Sequoias today are native to North America, more specifically to the Sierra Nevada in California; the first ones planted in Europe were planted in the 1850s. There are times, especially when I'm working in the vines and look up to see the sequoia standing before the east façade of the house, that the tree becomes a portal into the past: I imagine it being planted, and hear the clattering of horse-drawn carriages pulling up in the courtyard below, and the generations of workers and inhabitants strolling by, the builders, farmers, gardeners, growers, winemakers. The first part of the house was built over 400 years ago, in the early 1600s, with the largest section added in 1766. In moments like this, when its history becomes palpable to me, I am reminded that nobody is really the owner of this place. We are just temporary stewards. And like all those others, I will always be part of this place too. This feeling, of being both temporary and permanent, has become part of my everyday life, and it largely defines how I see myself and comprehend what I'm doing here.

Before pruning can begin, the trellis wires which hold the previous year's growth need to be detached. Left to their own devices, vines will go wild if not given boundaries. The new shoots of the year will grow in all directions, looking for anything they can find to wind tendrils around, climbing towards the sky. The trellis wires provide order and keep the shoots together. In the spring, when growth has sent the shoots out to a certain length, the wires on either side of each row of vines are lifted to bunch the shoots together, upright. The wires are held together along each row with metal clips, and it is these clips that must be removed before starting to prune, otherwise it would be difficult, sometimes impossible, to remove the pruned canes.

It is this, undoing the several thousand small metal clips, that kicks off my winter. Given the cool Loire Valley climate, it's best to prune as late as possible in winter, because the later you prune the later the first buds will break and the higher the chances of surviving the early spring frosts we have here from time to time. If you get a spring frost and the buds have already broken, the harvest is destroyed. So I try to start pruning in January or, at the very earliest, the second half of December. If you travel around the region in winter, you will notice that much pruning starts in late November or early December; pruning is so labour-intensive that many vineyards have to start this early in order to get the pruning done on time. This is the only vineyard activity in the winter that gives the winegrower a gentle but constant reminder that there's a deadline looming. Winter can be lived more slowly, but it is absolutely essential that all the vines are pruned before the sap starts rising and buds start breaking in the spring.

And so it is that in the first half of December, I can be found walking every row of the vineyard with a cloth bag slung around my shoulder, undoing metal clips and slipping them into the bag. I might even have a little smile on my face as I think about the bag, a remnant of my old life, bought on a trip to Havana. The iconic image of Ché Guevara, with its background of the Cuban lone-star flag, never failed to cause a raised eyebrow among my more conservative colleagues.

The protective calluses that build up on my fingertips from the repeated squeezing of metal clips and pulling at wires are a reminder that my body is an integral part of this work, and reacts to it independently. It doesn't need the brain for everything. Although it may not seem particularly stimulating, I love walking every row of the vineyard removing metal clips. What better job is there than going for a walk among vines? I will usually raise a hare, send it off like a bullet, and might see a roe deer or two nibbling on the grasses between the vines, that short glimpse before their ears alert to my presence and they prance away into the distance. On the luckiest of days, I will witness the hallucinogenic display of starlings, murmurating around me with the feverish rush of thousands of beating wings, tens of thousands of birds swirling and swooping in unison, the dance of an unknown choreographer.

Is it possible that a place can change the shape of time? Juliette and I are convinced of it. Before she came to the vineyard her life was in Paris, in the world of art. She spent her time in galleries and international art fairs in Paris and Dubai. Then she became a grower, a mother and smallholder. Sometimes, when we sit outside at the end of a day's work, we talk about how different this life is from our previous ones, and how it feels like we're living in a different time. Almost as though we inhabit another era.

Juliette's parents were farmers early in their married lives, and truly did live in another era. Albert, her father, came from the land. One of fifteen children in a family descended from generations of Normandy cattle and dairy farmers, he would regale me with stories of his youth. Sleeping five in a bed, head to toe. Breaking the layer of ice on the water of the bedroom wash basin in the morning to rinse his face. Finding a bloody, lifeless German soldier lying in a ditch after the Normandy landing and taking his helmet to use as a cooking pot.

Albert married Bernadette, a town girl from Deauville on the Côte Fleurie, and they got a small piece of the farm to make a go of on their own. Juliette was born there. But even then the life of a small family farm was getting harder. Working eighty-hour weeks, tending the cows, growing the hay to feed them, milking, raising children, and barely making ends meet. They loved that farm and held on for as long as they could. But by the time Juliette was five years old it was too much. Physically and mentally, they were at breaking point. The farm was sold and they uprooted from Normandy to the south of France, searching for a new life in Nice. Juliette grew up in this city with her distant memories of life on the farm, hearing her parents lament leaving it for the rest of their lives. She could never quite understand – it was difficult for her to reconcile the stories of hardship with this deep nostalgia. She would not understand until she came here to her farm, her land, her place.

Juliette knew from a very young age that art, particularly contemporary art, was her passion. If not spending her life making art, then she wanted to work with it some other way. She took a degree in Arts Administration and set off into that world, eventually ending up in Paris working for an agency that protected artists' copyrights. She then worked with a publishing house producing art books, and ended up managing the art of a wealthy collector.

At each rung of the ladder, her disillusion accelerated as she realised that sharks in the art world looked and swam the same as any other corporate world. When she told her parents that she was leaving her job, quitting Paris, turning from the life she had so wanted to live from her youth, to live instead on a vineyard in the countryside, in a crumbling old house with vines and woods, no neighbours in sight, Bernadette smiled at her gently and said, 'Ah! My dear child, finally you understand.'

Generally, the prevailing west wind breathes its way up the Loire Valley from the Atlantic, keeping things temperate here in Touraine. But in winter the wind can shift, raging from the north-east. On such days, I have to wear my full wet-weather gear with insulated, knee-high rubber boots. Tiny pellets of hail come down and the wind whips them against my face. They hit my skin with tiny pinpricks, and I can hear nothing but the wind and the rat-tat-tat of the hail bouncing off my rubber hood. In the dawn's grey glow, I am hunched over a seventy-year-old vine, pruning it carefully. Despite the weather, I need to focus. When I pruned this particular vine last year, I left what is called a *rappel*, a 'reminder' bud.

The approach to vine formation and, therefore, pruning, varies widely across the world. There are vines that are made to grow high and produce the fruit at shoulder level. There are vines that are pruned to grow without trellising, and others where a single long cane is left each year and attached to trellising. The shape, height and pruning method of a vineyard in Austria compared to Australia, or Chile compared to Cyprus, are completely different. Even across the wine-growing regions in France, pruning techniques vary. These variations are fascinating and have evolved over the centuries as people have learned which approach encourages each variety of grape to produce its best fruit, given the climate and soil of each place. In Vouvray, we prune to keep the vines low to the ground. This tends to be the technique in other cooler climates, where the grower is trying to keep the vines and the grapes as close to the warm earth as possible. The earth and stones take in the heat of the day and share it with the vines at night.

One of the challenges of this method is that every year, little by little, the vines rise further from the ground. At a certain point, the old vines start to 'outgrow' the trellising system. Our oldest vines, which were planted almost a hundred years ago, are too tall for the trellising wires to hold their shoots and fruit in place. They require special effort and care to ensure that the shoots, when in full force of growth, don't grow too loose and put themselves in danger of being broken off by a strong wind or a tractor ploughing. Even though old vines tend to be considerably less vigorous and produce fewer bunches than young vines, their fruit has a more complex aromatic palette.

When I stand before a ninety-year-old vine, I feel the same sense of awe and weight of history as I do when looking at the house or the sequoia. I try to picture the person who planted the vine, pruning it through infancy. I wonder what their life was like and how different, or not, were their thoughts, dreams and desires from my own? Did they know the horrors of the First World War trenches? Did they hear a young Édith Piaf crackling through the wireless during the Second World War? Crucially, what did their wine taste like? The other, more basic pleasure of looking at an old vine is its beauty. The complexity of its form, winding and gnarly, thick and knotted, and the texture of its bark, rough, furrowed and flowing. The sheer strength of its trunk, too, standing firm on roots that plunge deep into the earth. All these things make it worth spending time with an old vine. This is why keeping them as long as possible is important to me. And why the reminder bud is essential, helping me gently steer a plant towards its natural, ripe old age.

As I prune, occasionally there will be a shoot that pushes out lower down on the trunk than the existing 'arms' from which the year's fruiting shoots are sprouting. These lower shoots are considered to be 'suckers' and are usually pruned off, but if they are in the right position, you can prune them very short, leaving just the one bud. Then, the following year, if this bud has given off a good solid shoot in the growing season, that shoot can create a new 'arm', lower down on the vine. The older, higher arm is then cut with a handsaw, reducing the height of the vine and encouraging its longevity. This process requires increased attentiveness and more work and time than just pruning for the fruit, but it is an effective way of keeping the vines as low as possible over the years. It also has a Bonsai effect on the tree (it's hardly natural for a ninety-year-old vine to measure a mere 80 cm in height), encouraging its gnarled splendour.

I prune the reminder shoot down to three buds and dig through my wet-weather outfit for the handsaw in my pocket. I kneel before the vine, saw off the arm that the new shoot will replace, and move to the next vine, taking a few moments to evaluate what it needs. Any reminders? Any dead wood? How vigorous is the vine? How many spurs should I leave? How many buds? Bending down, I start cutting each cane from the previous year. With each cut, I make an almost unconscious decision based on an internal picture of what I want the vine to look like, and what I want to ask the vine to do for me.

Sometimes the work is so hard, the weather so unrelenting, the plants so resistant to your intentions, that it pushes you to despair. During one of our first dinners together, after they became friends as well as mentors, Vincent and Damien broached the subject. I was getting to the end of my courses at the Lycée and my diploma would soon be in hand. I would be able to terminate the lease on the vines and take them back after that year's harvest. Three couples sat around a winter table, the main course was finished, remnants of a rich boeuf bourguignon smudged on the plates, wine glasses were refilled. 'We've been wanting to talk to you about something,' Damien ventured.

The warmth, the sense of comfort and well-being brought on by the food and wine had me completely unaware of any potential uncomfortable turn the evening might take. 'At your service,' I replied, raising my glass.

Vincent dived in: 'Are you sure you want to do this?'

At first I had no idea what he was talking about. Opening another bottle? Skipping the cheese and going straight to dessert? I noticed Tania, Vincent's wife, exchanging an uncomfortable glance with Damien's wife, Coralie, who in turn looked at Juliette, who seemed as oblivious as I was.

We soon discovered that they had come to warn us. They wanted to tell us that the life of a winegrower was hard. That it was filled with daily challenges, and many frustrations. That the physicality of it could be debilitating, that making wine was one thing, but selling it was even more difficult. That we didn't really understand what we were getting ourselves into. They came from generations of winegrowers and farmers, were born into it, lived and breathed it. They saw, after spending time with us, that we came from very different worlds. They worried that our expectations of what life should be like, the comforts and rewards it should provide, might not be in line with the reality of where we were heading.

The discussion became strained, at points heated.

They were, in fact, trying to discourage me, or at least give me second thoughts about this project of mine, this new life that I was craving.

The evening ended, and although we parted on good terms, it left a bad taste in my mouth, a slight feeling of betrayal. It's only now, with years of lived experience, that I see they were right to say what they said. It came from a place of care and concern. They were trying to manage my expectations because they could see what had brought me here; the dreams and desires, the hope, and even fantasy. They knew that the life they were leading, the one I wanted to lead, was not the one I had in my head. The hardships were far more present, the satisfaction and moments of joy more subdued.

I was puzzled. Did they doubt Juliette and me? Maybe we didn't have what it took. Why, after all this time learning and spending time together, did Damien and Vincent choose this moment to warn us away?

It was only a few years ago that I finally understood what they were saying, and why they had said it. And this time it was my turn to play the devil's advocate: I was across from Antony, a man in his late forties, then living in Brussels. He had visited us several times over the previous year or so, in awe of what he saw as our lives. His eyes went round with wonder every time we walked the vines or descended into the cellar to taste the wine. I could almost see his project hatching in his head. He wanted to quit his business, pull up sticks, do what we had done. He would bring his wife and two teenage children to the Loire, and *live the dream* (as he put it). He was not the first person chasing dreams, and will not be the last to cross my path.

As we sat, tasting the latest vintage, he told me that he thought he'd found a vineyard to buy. His personal savings plus a bank loan would allow him to do it. He told me this with almost wild eyes, looking for me to tell him to pounce on the opportunity without hesitation. I suddenly had a flashback to that dinner table with Vincent and Damien, decades before, and saw myself then: wild-eyed, not wanting to hear what Vincent and Damien were trying to tell me. Yet, I found myself repeating the words they'd spoken to me. I could tell that Antony didn't want to hear it either. 'This may not help you,' I said. 'But here is my riddle for you: If I had known then what I know now, I probably would not have done it. So I'm very glad that I didn't know then what I know now.'

I've seen others give it a try and leave. Not realising the work it involves, the sacrifices, the piddly income. Antony will try and see if it works for him. I hope he finds his way. Or, at the very least, *learns* what he is looking for, because it is there, even if it leads to somewhere else.

The hail has turned to middling rain, but the wind is still wild. The water is running down my hood and blowing into my face. My gloves are sodden and, despite being completely encased by the rubber gear, I swear the water is still getting in. This is one of the rare moments when I ask myself whether I wouldn't be better off back in my old life, safely ensconced in a conference room with coffee or comfortably seated in an airline lounge. It has been over twenty years now, and every time the answer comes to me even more quickly, *No, never.*

Instead, I channel my inner child and imagine that I'm sailing solo in the Vendée Globe around-the-world race and am on deck fighting against the storm, or that I'm on a polar expedition and the dogs are struggling as the wind rips at my sealskin coat. It may sound odd but to me there really is something heroic about pruning a vineyard, rain or shine, hot or cold, sore and exhausted. It was one of many great surprises of entering this more physical life: the daily challenge, the regular pushing of my body towards its limits, the recognition that the mind and the body are so closely linked. It is too easy to forget all this, and whenever I see others out in the fields pruning, I salute them. Most of them are employees at another vineyard. Often, they are migrant workers doing the work that most people in France won't do anymore. To show up, day after day, week after week, pruning other people's vines for minimum wage, you must be a certain breed of hero.

When I am feeling mentally strong and motivated, a day's pruning is so much easier than when there's some anxiety gnawing at me. The energy feeds on itself. But I have learned that you can shift this feeling by relying on the body more than the mind, allowing it to build its rhythm, muscles flexing, lungs working harder. The body feeds the mind and tells it that this is exactly what you should be doing, at this very moment, quenching doubts. 'If you are not here, now, you are nowhere'.

And decades after I first read Walt Whitman's shimmering poem, 'I Sing the Body Electric', it was pruning that finally allowed me to understand it.

The thin red jellies within you or within me, the bones and the marrow in the bones,

The exquisite realisation of health;

O I say these are not the parts and poems of the body only, but of the soul,

O I say now these are the soul!

Early and mid-winter pruning holds a particular delight at midday: lunch. Working outside all morning in the vineyard, especially when it is cold and wet, generates a serious appetite. So at the stroke of twelve I head back to the house. I remember my days in Paris and London, having dinner after 9 p.m., usually out at restaurants after a few cocktails. It's hard to imagine now that our meal pattern resembles that of farmers everywhere: early lunch, early dinner, and often bed before nine in the winter.

Drying off, warming up, and sitting down at our heavy farmhouse table is untainted luxury. Dipping into a deep bowl of steaming homemade soup with a big chunk of bread fresh out of the oven. A block of goat's cheese from another nearby farm is almost rapturous. I have been a foodie for decades, but I have rarely, if ever, felt the degree of pure sensual pleasure from eating a good, simple meal as I do daily during pruning.

Working on the land, I've developed a more visceral understanding of eating and its relationship to the body's fatigue. The need for sustenance and energy becomes stark. Before, I remember skipping meals because a meeting was running over, or running to catch a flight and not having time to stop for food. Or even on purpose: to lose weight. After physical work in the open and breathing in the cold country air, skipping a meal would be equivalent to not charging a battery – you wouldn't be going anywhere. Of course, there are some winter days when the weather is so grim that the warm, sleepy feeling that sets in after lunch with Juliette makes me want to call it a day and curl up for a nap in front of the fire. And sometimes, we do just that.

Juliette, especially in winter, is my *louve blanche*. Years ago, early on in our relationship, I called her *my louve*, without knowing its meaning. I was joking, imitating a Frenchman with a strong accent saying *my love*, and she went along with it, never translating *louve*, just saying 'I like it when you say that.' And so, every now and then, I would repeat my silly French accent in English and call her *my louve*. When I discovered that *louve* is the feminine of *loup* (wolf), I knew why she liked what I said: I was calling her my she-wolf. But the word in French is much more gentle, and when I pronounce it aloud it conveys to me a certain sense of serenity and safety. In wolf packs, there is usually an alpha male and an alpha female, who mate for life, and are the only pair allowed to reproduce. In fact, the pack is essentially a family, guided and kept in order by the alpha pair. I don't know if I'm an alpha *loup*, 'primarily involved in foraging and food provision'. But if what I've read on the internet is true, Juliette is definitely the alpha *louve*, 'responsible for the care and defence of the pups, initiating and organising the movement of the pack, possessing great endurance and strength, excellent survival skills and strong family values, creating strong bonds with their mates and offspring.'

Although the grey wolf population is slowly increasing in southern and eastern France, there are as yet no *Canis lupus* living around the Loire Valley. If I had been a winegrower here in Touraine in the 1700s, however, wolves would most likely have been as much a worry as frost, drought and powdery mildew. A document uncovered from local archives describes 'man-eating wolves' between 1742 and 1755 that plunged Touraine into a 'continuous period of fear', with references to hundreds of attacks, many fatal. Although the idea of wolves returning excites me, the knowledge of what was reported adds a layer of complexity to the feelings I have while imagining the past; alone in the vineyard, taking a break, the possibility of wolves suddenly emerges from the woods, a ghost species bounding, as I stand there with a tiny pair of secateurs to protect me from becoming a meal for the *louve* and its pups. Despite the sense of time having stopped here, some things have changed and may yet change again.

One of the most attractive aspects of wine farming is the transformation. While most farmers harvest and sell their crops, usually for others to transform, a winegrower also turns the fruits of the earth into wine – a personal expression of the family and the place that will go out into the world and reach far beyond the grower. There is something universal about wine. It is consumed all over the world, and wherever it goes it can encourage conviviality, conversation, contemplation, relaxation, friendship, passion, maybe sex, perhaps even love. How could I not be excited by the birth of each new vintage?

I am not so naive as to ignore the dangers of alcohol. However, I do think that fine wine is different. Most people don't go out and buy premium wine to get plastered or to feed a habit. It's just too expensive. Not to say that nobody ever gets drunk on fine wine, but this is not its primary purpose. I always encourage new wine aficionados, especially young ones, to focus more on the aromatic complexity and richness of flavours. I try to transmit the tradition, history and culture surrounding wine and the satisfaction of figuring out the best wine to pair with different meals, rather than just concentrating on the warm comfort and euphoria caused by the alcohol. Though the latter is sometimes most welcome too.

While I am a fervent believer in the principle that wine is made largely in the vineyard, getting from ripe grapes to wine does of course involve human intervention. In the autumn, the grapes are pressed and the juice is funnelled into oak barrels in the cellar. By early winter, the juice from the year's harvest has been in the barrels for over a month. It is fermenting in the cool, dark cellar, striding out on the path to becoming wine. It will ferment through winter, evolving slowly, developing its aromatic complexity and becoming richer and deeper by the day. At the end of winter, it will then be ready for the finishing touches, already showing its personality and the full characteristics of the vintage.

I have a very different relationship to the cellar than I do to the vineyard. In the vineyard, I'm hands-on, tactile. I am constantly there: feeling the earth, the soil, the vines, watching, learning, working. In the cellar, I am much more of a spectator, with a *laissez-faire* approach. There is a great temptation to intervene in the cellar – when you study oenology you are taught countless ways to intervene. You learn about the scores of oenological products, additives and techniques that can be used to control, shape and 'correct' the wine – from things as basic as adding sugar to increase potential alcohol and roundness, or adding tartaric or other acid to decrease Ph, to adding calcium carbonate for the opposite effect (if the wine is too acidic owing to grapes not being ripe enough at harvest). At the very hi-tech end of the spectrum we find things like centrifugal technology, which, according to one of the machine's manufacturers, 'helps winemakers to produce at very economical rates – making more wine, in less time, with less effort and investment.' This pretty much sums up the whole 'modern' industrial approach to winemaking.

I use the term winemaking here as opposed to winegrowing on purpose. Today, the vast majority of wine is 'made' with increasingly less regard for the ecology of growing. The objective is to make wine in the most economically efficient way, which means producing wines that taste the same each year and cater to certain, predetermined consumer preferences. This aim also means an aversion to taking the time and exerting the effort required to make wine in an undoctored way. 'Natural' wine has become a real scene, with hipsters from Brooklyn to Bratislava jumping on board. But wine was natural before around 1950, which is to say for thousands of years.

To illustrate that this organic, low-intervention, 'natural' approach is not some passing fad, I like to ask sceptics if they have heard of a little estate in Burgundy called la Romanée Conti – one of the oldest and most prestigious producers of burgundy, with a reputation for making some of the best wine in the world. While 'best' is subjective, it is certainly among the most sought-after wine, with bottles selling for thousands of euros each, owing, unfortunately, to wine snobs and speculators with far too much money on their hands. It is hard to think of a less 'faddish' wine. It is undeniable that the wine is excellent and, by most definitions, can be considered natural. But Romanée Conti didn't jump on a bandwagon. They simply maintained their approach to winegrowing over the years, resisting the hi-tech promises that swept over the wine world in the second half of the twentieth century. The 'new' movement to create natural wines is really just going back to doing things the way they did them before the rush to industrialise wine that began after the Second World War.

These days it is almost impossible to imagine eight generations of anything, except an iPhone perhaps. We know, conceptually, that all our family trees eventually go back to the first humans. But how many people have any real connection to their family beyond a few generations? Through some old photos, perhaps? What would it be like if you could actually connect yourself to your great-great-great-grandfather or grandmother? What if you even had something in common, something you could both deeply relate to, and if they came to see you, could talk about together? What if you worked the same land, lived in the same place? It isn't that far-fetched in the wine country of France.

Julien Pinon walked into my life at his father's side. François Pinon was generation number seven, Julien number eight. I had known François for several years – before I came to the Loire, he had been one of my heroes. Scion of the Pinon clan and head of the eponymous winery founded in 1786, François had ventured out into the wide world in his youth, ending up in Tours as a psychologist. He was one of the first generations of children in French farming families to decide they weren't going to stay on the land; they wanted more from the world, more for their lives. The number of family-owned farms is in decline, in part, because of this. But when the call came, and his father needed him to take over the family estate, François made the choice to come back.

And as François himself advanced in age, he saw the time was coming for the next generation. But they had all dispersed, including Julien. Nobody was sure if any of them would be willing to come back. Would this finally be the moment, after over two hundred years, that the Pinon winery would no longer be owned and stewarded by a Pinon?

François introduced me to Julien, who was down for a few days for Christmas from Lille, where he worked as an urban planner. I was stopping by their place to borrow a piece of equipment and, as custom has it, a bottle was opened and we raised a toast to the holiday. I spoke to Julien only briefly then, impressed by his worldliness, thoughtfulness and obvious intelligence.

Afterwards, as I made my way home and the winter sky came down across the hills of vines like a dark shadow, I mulled over the fact that Julien was clearly torn over the decision he would have to make in the next few years. This wasn't the standard career decision that people make all the time. This was life or death. The life or death of the family home, and the land that had been worked by generations of his ancestors, the vines that had been planted by generations before him, the cellars dug into the limestone hillsides and the house and outbuildings built by the hands of his grandfathers. All of this heritage was balanced on his shoulders. This was an urbane, cosmopolitan young man, who had chosen city life, working in a profession that required cities – could he possibly drop it all and come back? As I arrived back at our place and saw our old house before me, I thought of the tears the Sauger children had cried when they had sold this place, how traumatic that had been for them. I truly feared for the Pinon family, and wondered if their story would be another one of change, of vanishing lives.

Le vin chante, the wine sings in winter. The French expression for the noise that wine makes while it is fermenting is exaggerated. This bubbling and fizzing is more like the sound you hear if you open a bottle of sparkling mineral water and put your ear to it. Yet I've learned that there is something to the song, and there are moments in winter when I go down into the barrel cellar, take out a bung, put my ear to the hole and listen as intently as I might to a new piece of music.

Very early on in my winegrowing adventure, I took the music analogy to heart and installed a primitive sound system in the cellar. Now music plays softly every day, non-stop all winter and spring in the cellar, accompanying the wine. It is always classical music, and each year I choose a different composer. The French have an expression for this type of eccentricity: *une folie douce,* a gentle madness. No, I haven't noticed any difference between the effect of each composer's music on the various vintages, and no I don't *really* believe that the music makes its mark on the wine. But I can't be absolutely sure that it doesn't either. The real point, though, is that it can't do any harm, and it's wonderful to descend into the pitch-black cellar and, before turning on the lights, hear strains of Bach or Chopin winding their way up out of the darkness, carrying with them the distinctive aromas of an active wine cellar.

In the entire winegrowing process, from beginning to end, from dormant bud to wine in the bottle, there are so many things that we *don't* know, that we can't fully explain with chemistry, biology or any other science. This is the magic of it, the alchemy of wine, so why not add a little touch of one's own *folie douce*?

My principal job in the cellar in winter is to listen to this song, which is to say to make sure that the yeast is doing its job. The singing is the result of fermentation: the yeast converting the sugars in the grape juice into alcohol and its by-product: bubbles of carbon dioxide. All wine starts as sparkling wine. When you taste a sample in full fermentation, it resembles fizzy grape juice. When fermentation stops, the bubbles slowly leave the wine as it sits and becomes 'still' – or, as is so often the case with wine vocabulary, the French word *tranquille* describes this better. One way to make sparkling wine is to simply bottle the fermenting wine, trapping the bubbles in the bottle.

Our wine is made in French oak barrels. Fermentation takes place in the barrels, where the wine stays on and ages for one year. Barrels are extraordinary objects. One of the preferred options on my list for what I want to do in my next life is become a *tonnelier*, a cooper or barrel-maker. Whenever I buy a new barrel, the delivery is a much-anticipated moment: as the truck comes crunching up the gravel driveway, I feel a *frisson*, like a child at Christmas. I unwrap it, admire it. I smell the wood and run the palms of my hands along its rounded body. Wine barrels in France are still made by expert artisans who take pride in their creations. Like the oak trees from which they are born, each one is unique. Solidity and strength ring out from a barrel. A good one can last years, even decades, if it is taken care of – a good thing since they cost over a thousand euros apiece.

While the earliest wine containers were probably clay amphoras, the Greek historian Heroditus wrote that the ancient Mesopotamians used palm-wood barrels to move Armenian wine along the Euphrates River to Babylon. The use of oak barrels to store and transport wine became common during the Roman Empire, when it was also noticed that the barrels imparted appealing properties to the wine stored in them. Thus, the ageing of wine in barrels took hold. The magic of oak barrels is that the wood is porous to just the right extent – not too much, which would cause the wine to oxidise and spoil, and not too little, which wouldn't allow enough exchange with oxygen for the softening, rounding effect this has on the wine. The flavour that the wine takes from the barrel depends on its age and level of 'toast' – every barrel gets toasted on the insides with fire when it is being made, which is done in different ways to release the wood's natural sugars and helps bring out the flavours.

Occasionally, I go down into the cellar not to work, but simply to sit on a barrel, feel the wood grain against my palms, and listen to the music for a few minutes. One morning, my young daughter, Célestine, asked, 'Are there angels in our wine cellar?' She'd seen me come back from the cellar and heard me speaking about '*la part des anges*' or 'the angel's share', which refers to the wine slowly evaporating through the porous wood. This vaporous wine, the saying goes, is wine for the angels. In fact, it's mostly alcohol and water in the wine that evaporates, which in turn helps concentrate the aromas and flavours, making a barrel-aged wine more intense and complex than its sister made in stainless-steel vats.

Alas, wine made and aged in barrels is an increasing rarity. Most wine, especially white wine, no longer sees the smooth, curved belly of an oak barrel. Only high-end fine wines get this privilege. Like so much in our world, using barrels is too expensive and it takes too much time and effort to raise wine this way. Any 'oakiness' in most wines today is put there with oak chips poured into the steel vats and left to soak. The slow exchange with oxygen is imitated by micro-oxygenation technology: basically, blowing bubbles into the wine. Is it the same? Of course not. Would the average wine drinker notice the difference? Mostly no. So why does it matter? Well, for me it matters because of the gratification I get by taking the time and making the effort to carry on a tradition and way of life that have value for me. I have great difficulty equating 'faster, easier and cheaper' with 'better'. So yes, my dear Célestine, there are angels in our cellar. They are there to take care of the wine for us, to measure the passing of time, the deepest secret in the magic of this place.

In my previous life I was never without a watch. I measured and weighed time carefully. It was the scaffolding surrounding all that I did. Appointments and deadlines, planes and trains, carefully tracked and dosed. I must have had five or six watches sitting around, most of them Swatch and the like, but two of them were quite beautiful, both gifts. I've kept them but never wear them. They're in a bowl on a chest of drawers in our bedroom. I walk past them every morning and every evening. Usually, I don't give them a glance or a second thought. Occasionally, though, I will pick one up and look at it, running a thumb across the smooth face, its hands forever stilled, unwound, flywheel frozen, stuck in some long-gone moment.

When I broke from pruning one day and sat on a fallen tree, sipping hot coffee, I looked down at my bare wrist and realised I'd not worn any watch for years. Decades even. Sitting on that log, I remembered my habit of winding them tight to keep them alive, the second hand sweeping feverishly. The vines are my measure of time now, the piles of cut canes punctuating the row in front of me. It's as though the two versions of time that I've known in my life are entirely different species. What time will you be here? How much time do we have? How much time does it take? Will I make it in time? How quickly can you get it done? Mostly, these questions no longer make much sense to me. What time will the clouds break and the rain fall? How much time will it take for the buds to burst and the fruit to set? Is it time for the leaves to unfurl? For the yeast to stop its work? How quickly can you get the grapes to grow? I don't know. You might as well ask: how quickly can you listen to Erik Satie's *Trois Gymnopédies*?

That said, being sure that fermentation is marching forwards in time, advancing without problems through the winter, is one of the few things that can keep me up at night. To be safe, I descend into the cellar about twice a week, unstop every barrel, one by one, and listen. If I hear the gentle fizzing, fermentation is fine.

Early on, fresh out of my viticulture and oenology degree programme, I would use a hydrometer. A hydrometer is a long, thin glass tube with a bulb at one end and gradations on the side that looks much more at home in a chemistry lab than an old wine cellar. You pop the hydrometer into the barrel and it floats in the wine, allowing you to take a reading by looking at the gradations. Based on the density of the liquid the instrument is floating in, the reading tells you how much sugar is left in the juice. You can take regular measurements with the hydrometer to make sure the juice is fermenting, and therefore measure how quickly fermentation is advancing. Alternatively, if you aren't in a hurry and are patient enough to let fermentation take its time, you can just put your ear to the barrel.

The danger of fermentation stopping, for whatever reason, is that the sitting sweet juice is prey to any number of bacteria and moulds, as well as oxygen, all of which can spoil the wine. In this case you have to inoculate the juice with cultured yeast, heat the juice, or add nitrogen to 'feed' the yeast. Or all of the above. In all of my fifteen vintages so far I have never had a blocked fermentation. Touch oak.

In all fairness, I do still use the hydrometer occasionally. About once a month, I will take a reading from each barrel and make a note of it. Part of the famous French penchant for paperwork requires winemakers to maintain a *Registre de Cave*, a Cellar Register, in which all cellar activities are recorded, including alcohol readings. I also like to compare how fermentation progresses for each vintage. Every year the speed at which the yeast does its supremely important work is different, if you let nature take its course and don't intervene. These differences between vintages are yet another part of the mystery and fascination of natural winegrowing.

This old house is built of the fine-grained, chalky limestone known as *tuffeau* in the region. The stone was formed 90 million years ago, when the region was under the ocean, from the skeleton deposits of marine organisms. *Tuffeau* is the proud face of Touraine. It has been the stone used for construction here for over a thousand years: from homes and castles, bridges and towers, to the majestic Tours cathedral.

This stone is as much a part of the people and place as vines are. Its hue, from chalk-white to light yellow, depending on the location of the quarry, contributes to the luminosity of the region. The Gallo-Romans used it to build the ramparts of Caesarodunum, the once-upon-a-time-city of Tours. The stones' pinnacle of glory came with the building of the stunning Renaissance castles of the Loire Valley, the French valley of the Kings. Chambord, Chenonceau, Amboise, Villandry and so many more of the most exquisite feats of human creativity were built with the blocks of *tuffeau* extracted from the hillsides ridging the Loire, Cher, Indre and Vienne Rivers. It was on and in these edifices that the porosity of *tuffeau,* which makes carving it relatively easy, was used to sublime ends. You could spend a lifetime studying the intricate, breathtaking and often surprising scenes and designs carved into the pillars, arches, hearthstones, windows and doorframes, spires and cornices of these chateaux. Because of the unique texture of the stone, stonemasons flourished as artists. In a sort of elegant utilitarian synergy, the caves created in the hillsides of Touraine by the forgotten stone quarriers became the future wine cellars of this land.

Much of the *tuffeau* in the vineyard buildings, the house and the cellar has been covered by a thin layer of *enduit,* a mix of clay, lime, sand and water, often used to protect the *tuffeau* facades in the seventeenth and eighteenth centuries. But the *enduit* has been coming off this house in parts over the years, revealing the ancient stone brought up from local hillsides, leaving me to ponder which of the local wine cellars provided the building blocks of the walls that shelter me daily.

In addition to monitoring fermentation during the winter, checking on barrel cleanliness also comes into play. When the barrels are filled, the level to which you fill them matters. You have to strike the balance between too much and too little. If the barrels are too full, when fermentation kicks in and the bubbling and fizzing reaches its zenith, the barrel will overflow. Too low, and much of the juice risks being exposed to the air (and thus oxidation) before fermentation begins. Filling barrels the old-fashioned way, without gauges and auto-stop triggers, is not an exact science, and sometimes we'll get a little overflow running down the sides of the barrel which needs to be rinsed away. Fermentation also creates a yeasty residue around the opening of the barrel that needs to be wiped off occasionally, as does the bung. You then top up the barrel to give the angels their share.

Because natural winegrowing, by definition, leaves much in the hands of nature, and nature can be capricious, the same vineyard will never produce the same number of grapes each year. In fact, the vicissitudes of nature can lead to significant variations in the number and size of the bunches one gets in any given year. So you don't know exactly how many barrels you will need for a vintage. Naturally, you try to have enough barrels on hand, in case there are good harvests. But this means that in some years it's likely that some barrels will be empty, and an empty barrel needs careful attention, too, so that bacteria or fungus don't grow inside it, potentially contaminating the barrel and the wine when it's next filled. While cleaning agents do exist, the age-old method of burning a small disc of solid sulphur inside the barrel, then stopping it up tightly, works perfectly. If you can get hold of discs made from volcanic sulphur, even better. Either way, sulphur is a natural disinfectant.

So, on some winter days, I can be found in the cellar suspending burning sulphur attached to pieces of wire in the empty barrels, locking the glowing morsel inside the centre of the barrel by putting a stopper in the hole. It can get very atmospheric, especially if the music in the background happens to be *Carmina Burana* or some equally evocative piece. The smell of burning sulphur and the music conjure images of Lucifer himself, with my Faust grinning, holding a candle, face dimly lit, distorted shadow curving up the cellar walls in the faint orange light of sulphur flames.

The only other task that brings me down into the cellar in winter is one that involves pure sensual pleasure: tasting. As grape juice becomes wine, witnessing the development and progression of aromas, textures and flavours is beguiling. Long before I head into the cellar to taste, I have already sampled the pure grape juice at the press before it goes into the barrels, so I have a sense-memory of the original. And in the course of the twelve months, when the juice turns into wine and takes on its personality in the barrels, I will taste from every barrel at least twenty times. It is a key part of the upward learning spiral of winemaking. I call it a learning spiral, as opposed to a learning curve, because although I have learned much about winegrowing since my first baby-steps, there are many times when I'm forced to question myself, and feel like I've learned it wrong or that what I believed to be true is not, 'spiralling' me down to an earlier stage in my understanding.

There is no talent involved in tasting wine. Some people may be genetically predisposed to being able to discern smells and tastes more readily than others. Mostly, though, it's all about desire, concentration and practice. If you really want to be able to taste a wine and think *notes of summer honey and baked quince, and isn't that a little lime blossom showing through, or wow, what a stony finish*, you can. I don't mean that you can fake it. You really can taste all these things. But you have to work a little on it. The world's great sommeliers spend years, even decades, smelling and tasting hundreds of different flowers and spices and foods in the world, expanding their library of the senses, creating sense-memories and olfactory connections that can be used to sniff and swirl thousands of wines. But you don't have to do that to make a start. Mostly, you just have to concentrate a little at the moment of tasting a wine, and probe your memory. Your sense-memory will already hold all the things you've ever tasted and smelled, and it's about plunging in to find comparisons. Think of all the different fruits you have tasted, the flowers you've smelled, the woods you've walked through, and start by trying to find just one or two of these memories on the nose of the wine.

On the other hand, you may just want to taste a wine and say, *Hmm that smells good and tastes yummy. I really like that.* Frankly, I think both approaches have merit and I wouldn't judge either.

In the end, wine is personal – you like a wine or you don't like a wine, or you like one more than another or one with a certain food or if you're in a certain mood. The essential thing is that you enjoy the experience. The rest is secondary. But the time that I have spent focusing on taste has reawakened all my senses. Before I settled on the vineyard, my life was intensely cerebral, sometimes physical, but tremendously lacking in the sensual. These days, however, I'll find myself unconsciously bringing any number of things up to my tongue to taste or to my nose for a whiff. Walking through a forest, I'll pull a few pine needles or leaves from a tree, roll them between my palms, and smell. I'll pick up a handful of soil and do the same. At the farmers' market, as I go down the stalls selecting my fruit and vegetables, I'll inevitably and discreetly lift one of each to my nose before filling my basket. Not only will doing this reveal something to me about the ripeness of the fruit or flavour of the vegetables but it also just makes the whole experience of food shopping richer. And while I have always enjoyed food as more than just fuel, it has now become a keen pleasure. This has drawn me into the kitchen and set me cooking with relish – what used to be a burden has become a pleasure, almost the highlight of my day.

Here at the farm, as with all of Vouvray, chenin blanc is the grape that is grown to make our wine. It's a delightful variety. Its juice provides an unrivalled complexity of aromas and flavours. Over time it goes from fizzy sweet grape juice to develop the primary and secondary aromas, unfolding across a range of smells that includes lemon or lime, grapefruit, pear, quince, passion fruit, lychee, chamomile, green tea, flint, baked apple, dried peach and apricot, beeswax, cream, and many more. All of this from the juice of a grape! I will find some of these flavours and aromas throughout the entire process, from juice to wine. Some will ebb and flow over the course of the months, and some will make their appearance at a fleeting moment, never to be found again in that vintage.

Of the hundreds of aromas and tastes that can be present in a wine, the grape variety plays the most important role. There are flavours that can be present in a cabernet sauvignon that you will simply never find in a chardonnay. But even within each grape variety there are still dozens of potential aromas and flavours. Which of these are present and which are the most pronounced in the finished wine depends on that most enigmatic of wine terms: *terroir*. There is no good English translation for this word. Most often it gets translated as *soil*, and although *terroir* does include soil type, it goes well beyond that. It includes the climate and microclimate of a winegrowing area, the average rainfall by season and temperatures of course, but also things like: is there a river nearby that throws off mist regularly? Is the vineyard surrounded by a forest that acts as a barrier to certain weather influences? Are there large hedgerows around the vineyard that compete with the vines for nutrients? It includes not only soil composition but demands to know things like depth of soil before the bedrock is reached, and what kind of rock the bedrock is. Chalk? Granite? Some winegrowers extend *terroir* to include the hand of man: the traditional vine cultivation techniques in the region, for example, and even the overall philosophy of the person or people growing the grapes.

For a winegrower, there is nothing more important than understanding his or her *terroir*. Over the years, it is regular tasting throughout winter and spring, and then tasting the finished wine as it will be in the bottle, that develops this understanding. Along these lines, in an effort to try to deepen my knowledge of our *terroir*, I separate the pressed juice in barrels by date of picking and according to which parcel of vineyard the grapes came from. So, for any given barrel in the cellar, written in chalk on the front of the barrel, there will for example be: *Sept 16, South*. I have been doing this for over fifteen vintages now, tasting from each barrel regularly and learning. The complexity of *terroir* is dense, and I am convinced now that the learning never ends. One thing I can say, though, is that there is something common in all our wines from this place, from all parcels and all vintages: a certain minerality that translates into stony, salty, sea breeze notes. It's also rare that we don't have some underlying tones of white fruit (apple, pear, quince) and some more exotic notes (lychee, pineapple).

There are also a few basic truths about the differences between the land parcels. The East parcel, for example, gets plenty of the cooler morning sun while the South and West parcels get baked in the afternoon heat. This generally gives the wine from the East a vibrant, livelier character with citrus notes more prevalent, whereas the South and West will tend to be riper and richer with notes leaning more towards things like honey and pineapple. The North parcel, on the other hand, gets the least sun and ripens the latest. Depending on when we pick them, the North-parcel grapes can have lip-puckering acidity, which can give the final blended wine an uncanny freshness. While it is true that we could make several wines each year, either based on parcels or different blends, I have found that the different characteristics of each parcel comes together to form an extremely well-balanced blend. This is why we make one still wine each year, a combination of all the parcels of land, and therefore the expression of our *terroir* for that particular year – although these underlying characteristics are always present, every year there are differences and surprises layered through, which is the magic of making natural wine.

Even as the sun rises, February can be very cold, often below zero, with a light dusting of snow running across the tops of the vines: white lines on black arms. When it's time to go out to prune, I get dressed for the weather: long underwear, wool socks, rugged work trousers with kneepads, a wool sweater, hat, padded leather gloves, insulated boots. Having cleaned my pruning shears when I came in at the end of the previous day's session, getting the dried sap off, I then oil them and rub the sharpening stone back and forth over the blade. Finally, I use a special mini wrench to tighten the pivoting mechanism where the blades are joined. This may all seem a bit theatrical but, believe me, when you are going to spend hour after hour cutting through thousands of dry, hardened vine canes, the state of your secateurs begins to take on religious significance. Even with my shears in tiptop shape, the last things I put on are an elastic wrist brace and an elbow brace to help take the repetitive muscle strain. Finally, I fill a thermos with coffee and head out to the East parcel where the first rays of the sun are setting the treetops aflame.

Outside in the woods, the winter has brought out the character of each different tree species: oak, pine, ash, chestnut, beech, elm, alder, acacia. Despite encouragement from many neighbours around us to 'manage' the forest, to implement a plan to cut and sell wood in rotation, we have decided to let it go. The only way we take advantage of the forests' resources is for firewood. A four-hundred-year-old house is not very ecologically sound unfortunately, especially when it comes to heating it in winter. In an effort to limit burning heating oil we have installed a number of wood stoves in the old fireplaces throughout the house.

To fuel them, we don't cut live trees from the forest but only cut dead trees or take ones that have fallen. The natural cycle of a forest provides us with more than enough to fuel the stoves each winter. We also make sure to leave plenty of fallen trees in the woods, since they play an important role in the overall ecosystem. As they die and decay, they provide a habitat for wildlife, such as the woodpeckers and bats, or the colonies of insects and microorganisms that help cycle nutrients back into the forest floor. These dead trees can be anywhere in the woods, surrounded by living ones, which means accessing them and getting the wood out can be tricky. But there is still a strong tradition of horse-logging here, and we make enthusiastic use of it with the father-and-daughter team, Philippe and Lou. To see a massive workhorse like Mascotte pulling a felled tree out of the woods into a clearing, where it can be easily cut and stacked, is utterly awe-inspiring. The might of the animal and the skill of the person guiding it, all set against the background of the forest, really is a sight to behold.

In between the woods and the vines, there's a strip of unused meadow where veteran apple trees and a quince were planted years ago, well before my time, and which go completely untended, continuing to provide enormous amounts of fruit each year. Every year in the orchard, we prune the newer trees, cut away dead wood, treat them with natural oils against fungal disease, hoe away the grass at their bases, mow the rest of the grass, and put out water for the birds to try to stop them eating the fruit for hydration. Even so, we get less fruit and have more insect problems, disease and tree mortality with these trees than the wilder, older specimens left untended in the meadow.

Every year, I choose a parcel of land to renew. This year, it's the East's turn. The renewal process is crucial to any old-vine vineyard like ours. While cultivated grapevines can live for over a hundred years, many do not. There are any number of reasons why a vine might die. They might be suffering from disease or old age or sadly might have been torn out or accidentally broken while the vineyard was being ploughed. This leads to rows where there are 'missing' vines that need to be replaced. In wine-growing regions around the world, vine nurseries have become big business and it's easy to order a wide variety of cloned plants of any of the local varieties. The most common approach to renewing a parcel of vines is to walk through it, counting how many plants are missing, and then order that number from the nursery. Then you bring in a mechanical backhoe, drive it down each row and dig a hole where there is a missing vine, and plant the new one. There is another method, age-old, called layering, that we use but which takes more time and effort.

The basic principle of layering is to use one of your vines as a mother vine, from which a new 'baby' vine can emerge next to it. Rather than pruning all the vine canes back in winter, you leave one long one on the mother vine. Then, in early spring, you dig a hole with a shovel or pickaxe and 'layer' the long cane into the ground, placing it in the hole down one side and then bending it so it comes up out of the other side of the hole. You fill the hole in, and the end of the cane that is now sticking out of the ground is your new vine. It needs to be protected and shaped, so you hammer in wooden stakes on either side and attach the vine upright to a stake with a piece of string. This new vine will first be nurtured by the older vine, but will gradually develop its own root system and independence. Several years later, this umbilical join to the mother vine can be cut, and it becomes an individual.

Replacing vines in this way, as opposed to buying a cloned plant selected for certain characteristics (I have done both and now only use layering) is rewarding because its 'mother' supports the new vine in the first years, so I don't need to fertilise and water it. It's also already adapted to the conditions of your *terroir* because it comes directly from it. For me, every time I prepare a vine for layering, it reinforces the sense that I am not just growing grapes to make wine to sell on the market, but am also a care-giver, nurturing the vines, helping them reproduce and grow. It strengthens my connection to these plants, this soil, and the ongoing story of this place.

The short winter days give me and Juliette more time to spend in the kitchen. Better yet, it's guilt-free time because having used every minute of daylight for work in the vines by nightfall we feel free to indulge in our love of cooking. Living in central France, we are spoiled for quality produce, both local and organic. Our village has an excellent greengrocer, an outdoor farmers' market once a week, and there are numerous local farms that sell direct. All this, combined with the fact that we grow a multitude of fruit and vegetables ourselves, means that our cup runneth over with healthy, tasty ingredients. Sourcing them, preparing them and creating a fine meal has become a treat like no other for us.

Sometimes, I think back to being twenty-four and running around the floor of the Sydney stock exchange, shouting at the top of my lungs, feverishly buying and selling shares. If someone had told me then that I'd become a person who relished being in a kitchen, spending hours cutting, chopping, baking, stewing and searing, I would have cringed. And I probably would have had the same reaction to the idea of spending all my winter days pruning vines. Where was that piece of me back then?

And where goes food, goes wine. At least for me it does. In fact, cooking without a glass of wine at hand just doesn't work for me. Not only does it make the moment even more convivial, but it gets the taste buds primed, whets the appetite, and helps me to focus on the best wine to pair with the dish we are preparing.

Each season has its food. Winter is all soups and stews, root vegetables and tubers and roasts. We also regularly pay homage to our Swiss and Savoyard friends by putting on a *fondue* or *raclette*. Sometimes, we'll go all the way and open a Riesling or Chignin with the dishes, but most often we improvise with one of our slightly off-dry Vouvrays or even a demi-sec, which have the acidity to complement the creaminess of the cheese with aplomb. These meals are a favourite with the kids. We don't use any of the fancy electric *raclette* ovens that you plug in and put in the centre of the table. We found a company that makes individual, old-fashioned gadgets that use candles to melt the cheese. The kids love the autonomy of having their own little candle-ovens in front of them, where they can coax and conjure up their meal.

I always felt that our children should be made familiar with wine from an early age, and they were all introduced to wine as early as three or four years old. Initially, we would simply let them smell the wine or dip a finger into the glass and pop it in their mouths. Later, as they grew up, we would do the same thing but suggest they concentrate on the aromas and flavours, telling us what they could recognise. Sometimes we would guide them. Could they smell or taste the pineapple in a wine made from very ripe chenin blanc? What about the smokiness of a fireplace in a syrah from the Côte Rôtie? When they put their nose into a burgundy, we might say things like, 'Remember the walk we took in the forest last weekend?', trying to evince the damp *sous bois* or forest floor that so often floats off a fine pinot noir. As they grow into adults, our children may or may not be wine lovers. They may choose not to drink it at all, but we want them to have the tools to enjoy wine to the fullest if they do continue drinking it. We don't want them to look at wine purely as an alcoholic beverage, but to understand what's behind it: the history, tradition, culture, and the sensuality that comes with loving wine.

Between pruning and layering, the winter days are quickly filled. This doesn't leave much time for another important winter activity, which is repairing the infrastructure of the vineyard. The end-stakes and end-row support wires, the trellising wires and the row-posts all need ongoing maintenance. There has been a move towards using plastic and metal posts, but we continue to use wooden ones. These are, of course, less durable. Over time the base of the posts, the part in the ground, rots away. Wires rust and break. As I go through a parcel pruning, I make a note of each of the broken or fragile posts, snapped or weakened wires that I will need to see to later.

Besides being more environmentally friendly than their plastic or metal counterparts, wooden posts, to my mind, are more aesthetically pleasing, too. They are made from surprisingly rough-hewn acacia. Each one is an individual – no two have exactly the same form, knots or grain pattern. Piling them onto the trailer to bring out into the parcels where replacements are needed, I usually tell myself that I should really be making my own posts from the acacia trees in our woods. I've been telling myself that for years. But the days are too full and time too short for this project – it's on my very long list of things I'd like to do when I find the time. A list that grows longer every year; I'm beginning to think that much of what is on it will never happen, or will be for the next generation of winegrowers here.

Out in the vineyard, raising a sledgehammer high and bringing it down on the head of a post, driving it further and further into the earth, I sometimes think of the legend of the steel-driving man John Henry. I picture the strength in his arms, the nobility of his body, all so much greater than mine, and hope that I don't meet a similar fate! But for someone who had spent so much time in meetings and in front of computer screens there is something exhilarating in the sheer violent physicality of spending a day wielding a sledgehammer, and in the complete and utter exhaustion it brings. The sound of every blow; metal meeting wood in a gunshot explosion that rings out across our little valley shouting to all and sundry: I am here!

But as the February days start to grow longer and the head of March becomes visible on the horizon, I must put posts and sledgehammers down to prioritise my secateurs for pruning. Every morning I'm in the vines earlier, until finally the first slicing cuts can be heard at 7 a.m. By mid-March, I will be pruning in a t-shirt, sweating by midday. This is also when I start feeling anxiety mixed with the excitement of seeing the last rows of vines to be pruned. With the end in sight, sometimes I wish winter on the vineyard would last forever. Winter is anticipation. The anticipation of spring. Anticipation of the heat and light of summer and, finally, harvest. And as is the case with so much in life, anticipation is often the best part. But, of course, life without a little anxiety about what's coming next would not be life at all. The inevitability of spring bears down and the ever more vociferous birds persist in reminding me. 'Hurry up,' they say, 'You'd better finish pruning soon, or you'll be caught off guard. The sap is rising, life is coming back into those gnarled old vines, and if you are not careful you will spend the entire spring running, trying to catch up.' I begin to feel like Daphné, fleeing desperately from Apollo, picking up pace, forcing my secateurs onwards, faster.

PRINTEMPS
SPRING

Life is all aims and purposes and intentions. All its roads are intended to go from A to B. If only someone were to dedicate their life to building a road beginning in the middle of a field and ending in the middle of another.

FERNANDO PESSOA, *The Book of Disquiet*

The fields have come alive, the vines are murmuring. On the first spring mornings, I am an animal coming out of hibernation, looking around at a new world frenetic with excitement and potential abundance. Everywhere there is something new: leaves, longer shoots, new colours and flowers, grasses growing lush and taller. For two months there is luxuriant, daily change all around me. Nature can feel static in winter, reserved, but the eruption of spring is a reminder that we are surrounded by energy, active, pulsating life, and that I am a part of it.

Les vignes pleurent, the vines are crying, is how the French describe the waking of the vineyard in the spring. As the days and nights grow warmer and the afternoon sun reaches gnarled trunks, the vines show the first clues of the new life to come. From each of the thousands of cuts my pruning shears made over winter come these 'tears', the sap rising in the vines to form a gleaming droplet on the end of every cut stub, sometimes running over and sending a tear or two back to the earth. English-speaking winemakers call this 'sap bleed', referring to the 'wounds' of the pruning cuts which haven't had time to heal yet. Despite the painful descriptions, the tears and blood, this is an important moment for the preparation of new growth. It primes the entire circulatory system of the plant, getting nutrients and minerals flowing and helping to ensure a smooth and even bud-break in the coming weeks.

The sap running from the vines is mostly water, minerals and sugar. As the first tears fall, I always go out and drink some of the droplets, scooping them into my mouth. The sap is tasteless, but in my mind I taste the earth, the plant and even the fruit that this liquid is setting out to nourish.

The tears of the vines coincide with bursts of yellow and violet all over the farm from the daffodils and crocuses: the most extroverted signs of spring. Over the first weeks of the new season, the hedgerows detonate with firework-sprays of bright-white wild plum blossom. The rows between the vines are carpeted with tiny wildflowers in blue, purple and pink, and speckled with buttercups and daisies. And then there is the mimosa: every year I try to witness its transformation but fail. It always seems as though I go to bed one night, when the entire massive shrub is green, and wake up the next day to it in full bloom, its delicate yellow blossom swaying in the breeze and visible from our bedroom window.

Spring is also the time for men and their machines. Fortunately for wine lovers around the world, there are more and more women with their machines, too. But today in France, behind the roaring and clacking, ripping and screeching, you will still almost exclusively find a man. It is one aspect of spring that I am not enamoured with. The economics of today's world make it impossible to perform all the spring work on a vineyard without machines. The number of people I would need to employ would make the farm unviable, and it isn't easy to find willing workers. So, in April, I become a mechanic and a tractor jockey.

Despite my resistance to getting on the tractor, I can't deny that machines have simplified work on the land and lightened the farmer's burden. When *Homo sapiens* stopped hunting and gathering and set out to farm some 12,000 years ago, it was inevitable that a great, endless quest was set in motion: to find relief from the repetitive, back-breaking work. While steam engines were used in farming in the nineteenth century to power threshing machines and even some primitive tractors, the watershed moment came in the 1920s when relatively cheap and compact gasoline-powered tractors became widely available. Vineyards presented a particular challenge, however, because the tractor needed to be able to fit through the narrow rows of vines, and it wasn't until the 1950s that tractors started to become a common sight in French vineyards.

Initially, tractors simply pulled the same tools that animals pulled through the rows of vines, ploughing the soil. But with the invention of hydraulics, tractors could also supply power to tools that could perform other tasks such as pre-pruning in winter, shoot- and leaf-thinning, mulching, mowing, trimming and spraying. The plethora of machines that winegrowers now have access to is mind-boggling. From very early on, I tried to step back and triage the nice-to-haves from the must-haves. The fact that our farm can be considered a 'micro-winery' helps; the larger the vineyard surface, the more difficult it is to limit the use of machines.

And so it is that, to the singing of the robins, the crowing of the cockerel and the tears of the vines, one spring morning I slide open the massive barn doors and reconnect the tractor battery, which has been kept charged over winter. I turn the key. The roar surprises me yet again, and the smell of diesel fumes reminds me to hurry, to limit the use of this man-made beast.

When I lived in the shelter of the urban world, before I ever used farm machinery, I was blissfully unaware of the remarkable fragility of human skin. Since I've become a farmer, frequent but unintentional meetings between limb and metal have taught me much about the vulnerability of the human body. The number of cuts, gashes and bruises I have sustained is countless. Fortunately (especially if I look at the winegrowers around me) I have not sustained more serious injuries. The machinery that we need to farm economically is heavy, sometimes violent, and very often more stubborn than one of its predecessors, the mule. But at least mules were environmentally friendly.

The first machine I attach to my tractor in the spring is a mulcher. When I pruned in the winter, I placed the cut canes in small piles running down the middle of every second row of the vineyard. To dispose of these canes, you can either burn them or mulch them. Many French vineyards still use cane burners, which are often just steel drums cut in half and put on a metal frame with wheels, like a primitive barbeque. As the pruners go through the rows, instead of putting the canes on the ground in a pile, they will throw it on the fire and move the burner with them as they go down the row.

The advantage of this technique, besides effectively disposing of the canes, is that the pruners can take a break and warm themselves by the fire every now and then. It also destroys many disease-causing bacteria and fungal spores that would otherwise stay in the vineyard and return in the spring. The downsides are, of course, that burning releases carbon dioxide, and the pruners, if they don't already smoke, become chain smokers of vine canes. But, in my view, the most important drawback of this approach is that you are depriving the vineyard of valuable nourishment.

The cycle of nature, with leaf fall or when twigs and branches break off a tree, is to restore this organic material to the soil. Rather than throwing this natural fertiliser away, I mulch the canes in the rows, giving energy back to the vines as the mulch decays. I am not a Luddite. I am just judicious about the use of machines and technology. In this case, it makes a lot of sense to me to use a mulcher, and before machines it would have been impossible to do this. So, in this case, hail to the spring and the men and their machines.

A mulching machine is essentially a destroyer: the Shiva of vineyard equipment. In fact, when you look up the English translation of the French word for it, *broyeur*, you generally read *crusher*, which makes me think of a Marvel villain. The crusher consists of a central axle to which heavy chunks of metal with sharpened edges (appropriately called *marteaux*, or hammers) are attached by hinges. All of this is encased in a steel chamber. When you hook up this brute to the hydraulic system at the back of your tractor and switch it on, the hammers spin around at tornado speed within the chamber. You pull this behind the tractor between the rows of vines. As it rolls over each pile of canes, it shatters them to smithereens, creating nice, bite-sized pieces for the earth to consume and digest at leisure.

It is poetic justice that as the destroyer does its job, it also gradually destroys itself. Of course, all machines wear out over time, but agricultural equipment is particularly vulnerable to damage and degeneration because the work it does is extremely difficult and involves a lot of force. On the mulcher, the hammers need to be sharpened regularly, and eventually replaced. The belts that drive the axle wear out or even snap. If the pile of canes you run over is too large, or those particular canes were from very vigorous vines and therefore unusually thick, the hammers and axle can jam and the belts will spin on unmoving drives, first throwing off smoke as they heat up and then burning right through. By the time the sharp smell of burning rubber meets your nose in the tractor cabin, it's too late.

Our vineyard happens to be filled with beautiful, large chunks of flint. Flint and metal are not kind bedfellows. Dings, tears, metal bent out of shape are the result. I can often be seen with my tools lying under the tractor or one of my machines, hammering, banging, tightening or loosening, replacing drive-belts, fixing an oil leak, and inevitably bruising or cutting some part of my body. But even in the tractor or fixing it, at least I am still out in the vineyard where something new and surprising always happens, like the time I was ploughing and noticed a raptor gliding above me with its tremendous wingspan outstretched against a powder-blue sky. It isn't unusual to see birds of prey here, soaring on the warm air rising from the vineyards, on the lookout for field mice or other small prey. But this one was huge. It might have been a black kite that was closer to the ground than usual, and clearly had something in its sights. I stopped the tractor and watched. Suddenly, the bird was falling from the sky, plunging towards the earth at a startling speed, coming closer to me at the same time, striking a few metres away, right next to the stump of an old vine where a lush cover crop was growing. In an instant I saw its prey in its clutches: a newborn hare, no more than a week or two old. At almost the very second the bird hit its prey, the hare's mother appeared and jumped on the raptor, punching at it with her forelegs. The bird panicked, let the leveret go and stared at the doe hare standing on her hind legs, glorious in defiance.

I wondered if there was going to be a confrontation, but the regal bird, having weighed the risks, took flight. The mother turned to me, or rather the growling machine, five times her height and hundreds of times her size. She moved closer and stood up on her hind legs, forelegs outstretched, ready to fight. Flabbergasted, I put the tractor into reverse and slowly backed off. When I was sufficiently far away, I shut off the engine and watched as she went to her offspring. She stayed for at least another minute before loping off into the woods. I got out of my tractor and walked to the spot where all this drama had unfolded. My fears were confirmed: her leveret was dead. Blood seeped out from a wound in his back. I crouched in front of the lifeless animal as a lump rose in my throat.

When you live on the land, people change with the seasons, too. And every spring, without fail, like the cherry trees and daffodils, Juliette blossoms. I watch her from a distance, see the lightness in her movement as she harvests some strawberries or redcurrants to make jam, brings a hundred bottles up from the cellar to prepare an order, sands and paints the winery door, or hangs the laundry outside to dry in the newfound warmth of spring sunlight. Every year at this time, even as we advance through the years, she seems to me to be more prepossessing and vibrant than ever.

Juliette is our connection to the community, to the people around us. Left to my own devices, I would probably disappear into the vineyard or cellar or woods, never to be seen again. She keeps the balance in our lives. She keeps us connected to the outside world. In spring, when the weather is fine, she'll suggest inviting friends over for a meal, and prepare for these feasts with such relish that her energy is contagious. I no longer feel that having friends over for meals is an anodyne event. It's a strong and significant thread in the weave of the fabric of our lives, and her attention to detail reflects this. Her tables are often works of art.

She will use the land around us to create the mood: flowers, boughs, leaves, stones and pinecones will go into creating an atmosphere on and around the table. She loves trawling through the regular antique and flea markets in the region, picking up wonderful old dining accoutrements for a pittance. Bowls and plates of Gien porcelain, silver salt and pepper shakers, linen napkins, all discarded, unwanted now in the world of IKEA. In the flea markets, you can buy a full set of antique silverware for the price of a cheap stainless-steel set from that Swedish mastodon.

Years ago, in the course of spring cleaning, I saw Juliette taking all of our stainless-steel cutlery out of the drawer in the kitchen and putting it into a box. I asked her what she was doing with it. 'I'm going to sell it on ebay,' she replied. 'From now on we are going to eat all our meals with silver cutlery.' Her mother had given her the old family silverware years ago, and it had sat, untouched, in a box in the dining-room cupboard ever since. 'Imagine,' Juliette smiled, 'my grandparents ate with these knives and forks.' But silver tarnishes in the dishwasher, so after every meal we load the plates but handwash and then dry the silverware. Occasionally, we sit together at the kitchen table, polishing the silver.

There are moments on the vineyard that have an ethereal quality. It happens when there are no visitors, no deliveries, no wine shipments. I will be out somewhere, doing a job, not thinking about anything in particular, when the breeze picks up and, for the slightest of moments, it's as if I'm breathing with the earth. It happened the other afternoon, while I was collecting vine canes – although most of them are mulched, fed back to the soil and the plants, I always gather several piles each season to dry out and use as kindling to light the woodburners in winter. I was kneeling to bind another pile, with the spring sun warming my back, when I felt the cool movement of air rustling the tall grass, slipping through the vine branches and leaves, washing through the trees. All at once, only for a moment, it seemed as if the earth was breathing, in and out. A cloud crossed the sun, the warmth left my back. I felt a distinct lightness. Sometime in the future, I thought, I'm going to be in the earth, and of it. This thought wasn't sombre. I wasn't scared. And although I'm agnostic and don't believe in any gods, there are times when, tending the vines, it does feel like I am doing god's work.

These moments are surprisingly fleeting, but I suppose that's what makes them special. Although many hours of my life are spent like this, alone with the vines, the farm is generally a busy place, with people coming and going throughout the seasons, all bound by the process of growing, pressing and making wine. One such seasonal visitor is Charlotte, who always arrives unannounced. She is slight, a little shy, and an extremely knowledgeable and competent barrel woman. She doesn't make barrels but knows *everything* about them, and if you are going to buy one, she's the person to ask.

Charlotte can tell you tales of the forest and the ancient oak trees that are at the origin of each barrel, and how staves hewn from those trees are weathered for years before being fired and toasted, then bent under the pressure of steel rings hammered around them to form the barrel. All of these are in her little bag of stories. She will take each part of the process out and elaborate as far as your questions will take her.

The trade of *tonnelier* is so exclusively masculine that when, a few years ago, the first female apprentice barrel-maker arrived on the scene it made national news, and a new word in the French lexicon found its place: *tonnelière*. Our winery is so small that it makes us quite insignificant to the overall sales of the *tonnellerie* that Charlotte works for. But it doesn't matter to her. She comes in the spring to be sure there has been juice in the barrels for months, and even though I don't buy barrels every year (when I do, I'll only buy one or two), she asks if we can go into the cellar and taste the wine from *her* barrels. She wants to understand how our wine meets the *barrique*, what that relationship is like, because every wine from every vineyard has a different interaction with her wood. She takes time with us and wants to be sure that I'm happy with the aromatic complexity, the smoothness, and the body that her barrels are imparting to the evolving juice.

There have been times when, ruminating on these subtleties with Charlotte in the dimly lit cellar, I feel like pinching myself. How lucky are we to be part of a microcosm on the planet in which the softness that ancient oak brings to a nascent wine is not just interesting but is genuinely considered to be important.

Natural wines really do like to be pampered – because they are very low in sulfites and other chemical interventions used to stabilise and preserve mass-produced wine, they are a bit more vulnerable. This means our wines are more prone to 'bottle shock', which can mute the aromas, and oxidation brought on by sudden changes in temperature and exposure to sunlight. So although wine orders come to us all year round, from France and around the globe, the most enjoyable and appropriate time for us to fulfil them is spring.

The process starts in the cellar, where the bottles of wine are stored in large steel crates in the best conditions of humidity, temperature and darkness. They are all brought out of the cellar using a forklift attached to the tractor, where they can be cleaned, labelled, boxed and secured to the wooden pallets that will take them to wherever in the world they are bound. We don't have well-lit, temperature-controlled warehouse space in which to do all of this. In fact, most of it is done outside, next to the winery, which is why early spring is ideal: it's already light enough to work by 7 a.m., but the cool mornings are comfortable for us to work in and don't inflict a temperature variation on the wine as it comes out of the cellar.

We don't have a bottle-washing machine, so on many a spring morning Juliette and I can be seen outside at sunrise, mugs of coffee or tea never far from reach, with washing mitts on our hands, wiping each bottle with lukewarm water. Once clean, Juliette takes up her position on the labelling machine and I take the capsuling machine, which attaches that little sleeve you find on the top of the neck of the bottle. 'Machine' is a grand word for either of these devices.

Bigger wineries really do have impressive machines, which take in large quantities of bottles on a conveyor belt, wash, dry, label, capsule them and even put them in a box for you. Our devices are rather more primitive. By hand, Juliette puts each bottle on a roller, presses on a foot pedal that spins the bottle and allows her to attach the label to the glass from an adhesive spool. She then hands me the bottle, and I slip an aluminium sleeve on the top of the bottle, then place the neck of the bottle into another small 'machine' which fastens the sleeve. Each bottle then goes into a cardboard box – six bottles in each – onto which the final task of writing the vintage falls to Juliette, whose handwriting is far more pleasing than my untidy scrawl. When a box is filled, I tape it shut and put it onto the pallet.

Of course, we could do more to automate this process. In the beginning it was not economically viable, but even though we could now probably afford to do it, it's become a habit that also fits with our idea of being artisans. It also gives Juliette and me time together, to chat, to listen to music.

The way we do things is clearly not the most profitable. Growing wine in general is not a very profitable endeavour anyway, and growing natural wine is even less so. Before I had even started my wine adventure, I was regaled by those in the know with the riddle: 'Do you know the secret of how to make a small fortune in the wine business?' 'Do tell,' I would respond. 'Start with a large fortune,' was the answer. You certainly do not get into making artisanal fine wine for the money. But if you do it right and are graced with a bit of luck, you can make a living from it, and have an authentic, rewarding life.

Perhaps the pull to this work is, for some, a form of gravity. The mass of the land drawing us into orbit. For this is what happened with Julien Pinon, who decided to leave his life in the city to take on the family farm. From Vouvray to Lille, and back again. The news spread throughout the village, from the bakery to the church, to the market stalls and the zinc-topped bar: Julien Pinon was moving back to the land. You could almost hear a collective sigh of relief from the winegrowers of Vouvray. For at least one more generation, this vineyard would remain in the family. No outside investors or businessmen to squeeze the bottom line at the expense of tradition. Oh the irony of it, as I too breathed the same sigh of relief – the American who had broken the line of tradition of *La Meslerie's* 400-year history. How was it, in the space of five years, I felt so close to this group of local winegrowers? Their families had been working this land for centuries. Why did I feel such relief that it wasn't an 'outsider' who would take over the Pinon estate?

I owe this sense of belonging to people like Vincent and Damien, and to Julien's father, François Pinon, who all took the risk of accepting me. People who were willing to test the water, to see what my intentions were, what lay behind my being here. People who took the time to come and see what I was doing, how I was working the land and tending to the juice from these grapes, and who gave advice and saw that I took it or, at the very least, sincerely appreciated it. People who invited me to their homes and showed me what life here could be like, if you respected certain tides, the ebbs and flows of history.

Spring stinks, for about a week. When we lay down the manure in the vines, the sharp, rotten-egg smell of hydrogen sulphide envelopes our hilltop. You get used to it, though, and I almost enjoy the smell now because it signifies natural energy returning to the soil. I know that our vines will be better for it. But I'm still glad that it only lasts a week.

We buy manure from nearby farms, where the animals are raised organically, and adapt the dosage depending on how the vines performed in the previous season. Some years we don't use any manure at all, if we feel the vines have been vigorous enough for a few years running. The manure is a mix of cow, horse and poultry dung that is heated, composted and pressed into pellets. Unlike chemical fertilisers, it doesn't just deliver the minerals directly to the plants, it also boosts microbiological activity in the soil, which in turn improves natural composting and the creation of humus. Humus is critical to soil quality. It retains nutrients and moisture, traps the oxygen essential for root development, and prevents erosion by 'glueing' soil particles together.

Ideally, I would like to raise animals alongside the vines. It is one of the to-do things on my list of projects. I wouldn't need to buy in the manure, and it would allow us to become a more self-sufficient farm with a circular agricultural approach. Family farms used to be like this, but specialisation and industrialisation since the end of the Second World War have picked this apart. Fortunately, there are enough organic dairy and poultry farms nearby to support quality manure production.

But among the uncomfortable truths in the wine world, the one that bothers me the most is that the large majority of vineyards the world over are doused regularly with poison. Winegrowers are the Kings and Queens of Roundup. Monsanto loves them. Of all the different types of farming, viticulture is responsible for one of the most intensive uses of chemical herbicide per square metre of land.

Why are farmers obsessed with killing anything that grows in their fields other than their principal crop? Because other growth reduces yields and increases costs. It comes down to pressure to produce as much as possible per square metre, at the lowest cost. This is perhaps a laudable goal. It is what makes modern man modern. Unfortunately, as humanity is now learning, albeit very slowly and much to its chagrin, there is often a terrible price to pay for this form of 'progress'. Intensive use of synthetic herbicides not only destroys the natural fertility of the soil, poisons the water table and rivers, and severely damages biodiversity, but it also damages the health of those humans directly and indirectly exposed to it. It should not be surprising that chemicals which have been explicitly designed to kill organic matter – plants, fungus and insects – are a bad thing for humans to be in contact with. If you go into the agrochemical storerooms of 'conventional' winegrowers, you will find piles of bags on the floor and shelves brimming with plastic containers full of these chemicals.

I confess to fantasies of killing, too. Killing everything that grows in the vineyard, apart from the vines. Especially in the spring, when I start to see double, because alongside the rows of vines there are rows of grasses and other species of plant (most people call these 'weeds') that live here, too, and if I don't do something about them the vines will be totally overgrown.

All agriculture is about taming nature, getting it to do what you want. Even when you're farming like us – letting most things be, having the utmost respect for the forces and balance of nature, trying to intervene as little as possible – still, if you don't do *enough*, nature will decide for you. And you can be sure that nature won't do what you want. Nature is all about grabbing the opportunity to grow. Nature doesn't give a toss about me wanting perfectly ripe, sweet fruit that is easy to harvest. If I don't act, by the end of the season I won't even be able to see my vines, let alone have much fruit. The plants between the vines will grow and grow until it gets very difficult to harvest, and the massive competition for nutrients and water among the plants will make the vines struggle to produce bunches of grapes. The fruit will suffer further because the growth surrounding the vines will trap more humidity around the grapes, creating the perfect conditions for fungal disease (mostly mildew) that rot the fruit.

The purveyors of poison will tell you to kill all these species of plants that compete with the vines. Spray here, spray there. But nature is tenacious, and you will have to kill them again and again, year after year, until the soil is dead. As is often the case in these situations, the circle turns vicious. Once you have killed the life in your soil (poisoning the 'weeds' also kills the essential worms, insects and microfauna), you will need to put down loads of chemical fertiliser to nourish the vines because your soil can't do the job for you anymore. It's a sad cycle to see. When you walk in a vineyard, look at the ground; it can be a moonscape, barren of plant life, the soil poisoned to death.

In Europe, there are now many safety regulations around these products: secure storerooms, labels clearly explaining the toxicity to humans and danger to the environment, the protective clothing and masks to be worn when using them, the required quality of the air filters in the tractor cabin when spraying, and so on. Yet we spray these concoctions multiple times all over the crops that we and our children will later eat and drink. Many farmers – in private – will admit that they don't like the way they work, and some will not even eat the food that they produce. But they do it because, economically, they have no choice. And I understand that. The market demands it. But what is the market if not you and me?

As more and more scientific studies hammer down the solid doors to political power that the chemical companies have built, more products are being restricted and even banned. All too late for people like Jackie, Damien's father.

I remember that terrible day so clearly. Damien told me over the phone, in a small voice, *Cancer.* And just like that, my dear friend was now the head of the family business, at the much-too-young age of thirty. His father's funeral was at the beautiful Church of Saint-Denis. Dating from the twelfth century, in the town of Amboise, overlooking a majestic and peaceful Loire river, the church was filled to the brim with multiple generations, mostly winegrowing families. The solidarity was tangible. Although I sometimes felt like an imposter, here I was, welcomed and accepted alongside local families grieving for a fellow winegrower.

As is the custom, I waited in line to approach the coffin, where a receptacle was filled with holy water. Inside was a small sceptre, and when it was my turn I picked it up and shook it at the coffin, projecting drops of the holy water onto the shiny, polished wood. I took a moment to thank the man who had left us, and to tell him how grateful I was to know him and how thankful I was to have his son as a friend.

Even after knowing all we now know about pesticides and herbicides, it is still easier to simply spray everywhere on the ground, under and in between the rows of vines. But I don't. Instead, I choose mechanical tools. Vines, however, are fragile. Especially old ones, and it therefore becomes highly precarious dragging heavy steel tillers with teeth, blades, discs and other metal protrusions through the narrow alleys to break up the soil and bury the spring growth competing with the vines. It is all too easy to accidentally hit vines along the way, breaking off arms or snapping the entire vine from its roots. Much more difficult than even this is to till under and in between the individual vines in their rows. This requires different tools, a more delicate approach. For this, we invite Mascotte to the farm, accompanied by his human partners, Phillipe and Lou.

Le Comtois, le Breton, le Percheron. These names ring out in my imagination, evoking a time when we were closer to the land and to the animals we share it with. *Le Comtois, le Breton, le Percheron* are just three of a number of breeds of work horses that have worked the soil of France for centuries. Mascotte is a *Comtois*, and while each type of horse has its specific traits, they are all strong, majestic animals that were once essential. The horses that carried people on their backs, the breeds that we know as horses today, were only a very small proportion of all the equines that accompanied us in days gone by. Many more pulled ploughs and carts, carriages and trams. They powered the mills that ground our grain and the pumps that brought water to our crops. They transported our goods and us with them. They were buses, trains and automobiles, tractors and engines. They even fought our wars, carrying us on their backs or pulling cannons to the battlefields.

In the early twentieth century, there were nearly as many traffic jams as today, but the protagonists just looked (and smelled) very different. In 1930s France, there were over three million horses. Today, there are just a few hundred thousand. Electricity and the internal combustion engine led to the decline of horse numbers in cities. And although the French countryside held on to their horses longer, affordable tractors soon saw them off. They are making a small comeback in the natural farming sector in France these days, but this remains anecdotal and it's hard to imagine it ever being much more than that.

Working with horses takes more time and skill than working with a tractor. These skills are hard to teach and even harder to find – you have to spend hours behind a horse in the rain or raging sun, tilling the earth. You have to love your horses, and be at one with them. When I watch Philippe and Lou working between our vines with Mascotte, I see a connection between them and the horse that I never knew existed. They communicate in ways too subtle to hear, almost invisible to the untrained eye.

Philippe and I both came to the land late. Our paths were different, but he also moved from the city. He tells me that the young people he teaches to work the soil with horses find jobs immediately, and usually head off to Bordeaux or Burgundy, where the prices for those wines allow the winegrowers to invest in this 'luxury' of horses. Some of the most famous Grand Cru wines in the world now come from land tilled this way, which means the art of the blacksmith has also been revived in parts of France, hammering out horseshoes and the plough blades once again.

Despite this resurgence, I know that it will remain on the fringes of global agriculture. I don't think this makes it any less important. Watching the horses plays tricks in my mind, drawing out memories of things that I never lived through, painting *tableaux* of the early twentieth century in my thoughts. I enjoy the looks of wonder and awe in the eyes of both children and adults as they watch Mascotte pulling the tiller up one of the rows of vines, muscles bulging, glistening with perspiration, head rising and falling with each step, Philippe guiding the steel plough behind. Time stands still for everyone. The past and the present meet. And I know that there is value, too, in contributing to the continuity of *la traction animale*.

On the parcels where we can't use horses, I can be found sitting on my own iron horse, pulling the mechanical hoe, which extends and retracts around each vine, scraping the growth away. It's a slow process. I advance at less than 2 mph, constantly looking over my shoulder to make sure the hoe is working smoothly and that I'm not damaging any vines. Inevitably, I always break some. Each time, I stop the tractor and, with a heavy heart, climb down and pick up the limp plant and remove it from the vineyard.

With experience, you get better at minimising the damage the tools do. You develop a feel for the movement of the blades and get better at identifying the best moment to enter the vineyard. The state of the earth needs to be just right. If it's too moist, the tools will jam up with the wet clay; if it's too dry they won't be able to cut into the soil to displace it. In either case, the work will be shabby and the damage to vines will increase. In all cases, the hoe needs to be adjusted regularly to account for different gradients in different land parcels, and to minimise hitting vines. The perfectionist in me is never completely happy with the final results. I simply can't achieve the precision that Mascotte does with Lou or Philippe, in sync with each other, in sync with the land.

Between tilling, while I was sitting with Philippe in the shade of the Judas tree that leans into the East parcel, taking a break, my phone pinged. I resisted at first. The tree demanded all my attention, full blossom, a shower of deep pink flowers. The bees were having a field day.

Our three children always loved this tree. They climbed in it, played under it, sang in it, even fell from it sometimes. At one point, we tied a rope to an old tyre and hung it from one of the high branches. The children would shriek with frenzied excitement as they swung and spun out over the edge of the vineyard. Today, the tyre has gone but the tree remains, bigger, stronger and more beautiful than ever. My phone pinged again.

The French name for the tree: *l'arbre de Judée*, or Judea tree, reflects its eastern Mediterranean origins. The English name for the tree is less innocent – I had read that the exact origin was somewhat vague, but linked to a belief that Judas Iscariot, after his notorious betrayal of Jesus, had hanged himself from a tree of this species. It may also be a case of the name being lost in translation from the French, but it stuck, and grew. Some say that the tree's flowers were originally white but, after Judas hanged himself, they bloomed the colour of blood. There are other, less commonly known or used names in English for the Judas tree. One of them is the 'Love Tree,' on account of its heart-shaped leaves.

Philippe started tending to blisters that were developing on his hands from the friction with the plough's handles, while I was rubbing an aching knee. We were silent. Mascotte snorted. My phone pinged again, and I finally took it out of my pocket to read a series of messages and photos sent by a friend from my previous life. He was having a business lunch at the Ritz in London. The first picture was of a page of the wine list – ours was on it, priced at over a hundred pounds a bottle! The second picture was of a sommelier, dressed in black and grey livery with gold buttons, pouring a bottle of our wine. In the background, I could see the ornate chandeliers and wall of mirrors, and the rows of tables with the deep-pink upholstered chair back-rests, gathered like a branch of flowers from the Judas tree.

Pride? Hubris? Self-satisfaction? A sated craving for recognition? I felt a little of them all, seeing that bottle being poured at that revered establishment, despite having sold that same bottle for less than the price of a single glass at the Ritz.

But as I looked around me, at the animal, at the man sitting on the ground, at the rows of vines stretching down the hill, I couldn't make the connection. I just couldn't find the threads that bound what I was seeing in front of me with what I was looking at on my phone.

In the early days on the vineyard, I still had that craving for recognition: that need for realised ambition, acknowledged achievement. But over time, this land helped me shed some of this vanity. Like the traitor hanging from the tree, these emotions were fading away with time here in these fields. We used to enter our wine in competitions for medals and in tastings to try to be selected for wine guides or magazine articles. There are hundreds of these wine beauty contests every year, and the buzz from winning, from being selected and showcased, was heady, as low as the downer of missing out, judged not quite good enough. I no longer pursue these accolades. I don't bother responding to requests for samples for magazine competitions.

I passed my phone to Philippe. He looked at the pictures, shook his head and stood up, stretching his massive, muscular frame. 'Let's just do the work,' he said, smiling.

There are mornings in the spring when dew settles like crystal filigree across the vineyard. Hundreds of spider webs stretched between vine rows glisten with dewdrops in the dawn sun. A tableau of delicate lace. The persistence and industriousness of these arachnids is astounding. Even when I've been through the vines the previous day with the tractor, the webs will be back the next morning. These spider-worlds don't appear in the vineyards that are doused in herbicides and pesticides.

When we moved here and I first walked through these rows of vines, almost trembling with anticipation, I was blissfully ignorant. The rental agreement the Saugers had signed with a local winemaker had not yet expired, which meant these vines were not under my care. But I wasn't a farmer back then, and what I failed to grasp was the terrible meaning of what I *didn't* see in the vines. The person renting the vines had farmed chemically for several decades, and had I known then what I now know, I would have been aghast at the lack of plant life, the dearth of insects, the paucity of any wildlife. Fortunately, nature is tenacious and bounces back with uncanny stealth, speed and beauty, if we give it the chance.

In a vineyard, take away the chemical herbicides and pesticides, keep tilling to a minimum, and life returns in a flash, repairing and regenerating. After the Sauger tenancy expired, I took back the vineyard and witnessed this rebirth in just a few years. It was nothing short of miraculous. From my courses on soil composition, I knew at least some of what was going on invisibly: the revival of up to a billion bacteria in a single gram of earth, the restoration of galaxies of fungi, algae, protozoa, actinomycetes, all the microbes that are critical for healthy soil structure and nutrient and water recycling, and which had been decimated by chemicals. But it was what was visible to the human eye that left me most in awe. Soil that looked barren and suffocated, lifeless, with nothing growing between the rows of vines, all burned away by glyphosate, gradually became multiple shades of green. A variety of wild grasses grew year-round, peppered with a spectacle of wildflowers in the spring and summer; it felt like watching those time-lapse images of a flower blooming in the space of seconds.

Although this regeneration took many seasons, each year, from the beginning, there were new signs. One of the signs that made me almost ecstatic was the reappearance of the tell-tale tiny mounds of curly earth: earthworm castings. Tentative at first, a spot here and there, after several years they appeared everywhere, like the bubbles of deep-sea divers breaking the surface, signs of the bustling life below. The importance of the simple earthworm to soil health is equal in magnitude to that of bees to an apple orchard. Earthworms are natural fertilising machines. They consume plant debris and soil, their digestive systems concentrate the organic and mineral constituents, and they cast off fertiliser, both through their tunnels and on the surface of the land. Meanwhile, their continuous burrowing and channelling aerate the soil and improve drainage. Their casts on the surface provide constant regeneration of depleted or eroded topsoil. And as the soil in the vines built up its supply of nourishing organic matter, I noticed changes in the soil itself, its colour and smell. It became darker, a deeper brown, and when I picked up a mound of earth and crushed it in my hands it released that heady, earthy smell, like a full-bodied red wine, whereas before it had had a slightly sour, metallic note.

With this came ever more frequent encounters with all varieties of insects. In an unbalanced ecosystem, some vineyard insects, such as leafhoppers, mealybugs, grape-berry moths and aphids, have a negative impact on vine growth and the harvest. But there are insects and other organisms, including ladybirds, dragonflies, ants, birds and bats, that feed on these harmful ones and do no damage to the vineyards. When the ecosystem in the vineyard is healthy, there is no need for insecticides. A diverse undergrowth in the vineyard creates a habitat for insects, which in turn boosts the diversity of arachnids and birds and bats. And all of these, the prey and the predators, helpful and harmful, provide organic material to the vines in life (when they excrete) and in death (when they return to the earth).

Of the larger members of this delicate and beautiful ecosystem surrounding our home, bats are particularly interesting to me. On warm spring and summer evenings it is always a special moment to be sitting outside at first flight, when the bats start swooping over the vines in the half-light, diving with frantic perfection. In France, as in many other countries, bats are a protected species. They are the only flying mammals on earth, and their presence in any landscape indicates the health of a place. Over the years, we've noticed that bringing the vineyard and the farm back into balance has meant an increase in the population of bats.

My ever-growing conviction that leaving nature to do its own thing as much as possible has extended to how I think about soil. Because we cannot see and experience life below ground, it's easy to take it for granted. But the vigour of plant growth every spring in the rows between vines tells me of the tenacity of life. Given space to thrive, life returns to the soil, as do the worms, spiders and other insects that depend on this growth.

This creates a dilemma for me. Even though this growth between and below the vines can quickly dominate, impacting our harvest, it makes me increasingly hesitant to till between the rows. These are not grasses and weeds to me anymore; they are a cover crop that every year becomes an indispensable part of a broader balance throughout the vineyard. Tilling not only destroys the cover growth but also disturbs the top layer of soil, home to numerous worms, microbes and insects. Along with spiders and bats, ant colonies have returned to the vineyard in abundance, and driving a tilling blade through an ants' nest feels tragic to me. The months of careful nest-building are lost in a moment. This year, I would like to leave even more of the cover crop between vines, tilling less, mowing them occasionally, only when the vines risk being overgrown. This will take a new restraint and judgement on my part. And of course I won't get things right, searching for that balance between biodiversity and livelihood. But we have to start somewhere.

Industrial winegrowers, who use herbicides and pesticides liberally, tend to call vineyards that look like mine *pas très propre, not very clean*. This desire for cleanliness and order expresses something about our relationship with the wild. Why are hedgerows geometrically trimmed? Why do lawns have to be cut regularly so as not to become unruly? Why are fallen trees sawn into fire logs rather than left to become a habitat?

One day, walking in the vineyard maybe ten years into this regeneration, I stopped to pick a wild strawberry growing at the foot of a vine, and popped its juicy sweetness into my mouth. I distinctly remember thinking that this restoration, this renaissance, that I had witnessed all around me had also, in so many ways, occurred within me too.

The horse-drawn tiller. The tractor-drawn, spring-mounted retracting tiller. The Egretier manual intercep (affectionately known as 'the rack' for the torture it inflicts on its operator). These are our tools.

The old-fashioned Egretier is no longer manufactured but many natural winegrowers swear by it. For several years, we used it in place of the spring-mounted retracting tiller. We now generally use it only in small plots of young vines we've planted. Not surprisingly, young vines (up to about five years old) are very fragile, even more so than old vines, so using the auto-retracting tiller is impossible because it would damage the young plants. If you don't want to use herbicide, this means you either hoe by hand, or use the dreaded rack.

I tried hand-hoeing for a few years, but this is so strenuous that I wouldn't recommend it to anyone older than thirty, and even then, only to those who have the muscle and cardio fitness of a professional athlete. Hoeing a few fifty-metre rows of vines by hand is fun and a great workout. Hoeing dozens of these rows will need a lifetime subscription to a chiropractor or physiotherapist. I am exaggerating, a little. But the physicality of natural farming shouldn't be underestimated. It's something I think of more and more as time glides on. In days of yore, the many children born in the family would join in and, over time, the physical burden would be spread and passed on, generation to generation. Children have many more choices and opportunities for their future today, which is, I think, a positive thing. It does however contribute to the downward trend of small farms being passed on and staying in the family.

For now, I am still able to use the old Egretier. The machine itself is a marvel. With its chaos of springs and levers, it looks like something dreamed up by Leonardo Da Vinci and the cartoonist Rube Goldberg. The way it works, however, is quite straightforward, drawn by tractor, with someone walking behind to work the blades. On our farm, Juliette drives the tractor while I walk along behind, grabbing the levers, pushing and pulling them to action the blades which go in and out, around and under each vine, turning over the earth, taking off a shallow layer and the growth with it. Juliette drives slowly; any faster and the Egretier blades would start ripping out vines, and its operator would not be able to push and pull and grab fast enough. I have seen the shaking heads from some of the 'conventional' growers around here as they drive by, seeing me crouched over the levers, sometimes struggling to get the blades into the earth and back in time to spare the vine.

The old-timer that I bought the Egretier from had seen the transition of his family's farm from natural farming to chemical farming. When he led me to the Egretier, tucked away in the far corner of one of his hangars, covered in a decade of dust, he told me that no one had ever thought of trying to sell it. Who would want this relic? I knew he was chuckling to himself at my folly, but soon after we bought it, when I was working on a parcel between vines, he showed up and watched from a distance. We didn't talk, as I was too busy pushing and pulling levers, but before he left I could swear that I saw a ripple of regret move across his face. Perhaps he was tired of the productivity and cost logic that had caught up with him, nostalgic for the simplicity of the old ways.

With Juliette driving the tractor and me walking behind, we get into a rhythm working together. At the end of each row, I let go of the machine and lean against the end post to rest as Juliette makes the U-turn with the tractor to go into the next one As she turns and comes back towards me, we usually exchange an acknowledgement, a smile, a wink, a grimace. Say what you will about the folly of using an Egretier, but at these moments I feel closer to my wife than ever. My tractor-driving muse.

Each morning, our beehives rumble louder with activity. The warmer spring days are a signal that soon there will be blossom. All winter, the bees have been huddled around the queen inside the hive, shivering and vigorously flapping their tiny wings to generate the heat that they and the queen need to survive. They nourish themselves on the honey collected in warmer months.

Very soon after we moved here, we discovered five beehives lined up in the woods near the East parcel. Or rather, we noticed an old man dressed in a stained, worn-out beekeeper's suit waddling off into the woods. We followed, to see what was going on, and learned that the man's name was Pierre. He was a local beekeeper that the previous owners had hired to take care of their hives. The agreement was simple: Pierre gave them part of the harvest and kept the rest as payment.

Pierre has since retired but a young, enterprising beekeeper, Jean-Christophe, soon followed in his footsteps, and the honey we get every year is unlike any other I have ever tasted. I've always enjoyed honey, but before Pierre and Jean-Christophe arrived in my life, it was just a golden liquid that came out of jars, usually mini-jars on hotel breakfast tables. Since being here, I have learned that it comes in a whole spectrum of colours – from white, yellow and gold to the deep, dark brown that is almost solid.

Real honey is relatively expensive because it reflects the work and craftsmanship behind it. As with so many things, the honey most of us know today is industrial. It is pasteurised and filtered, which destroys most of the vitamins, minerals and antioxidant properties of natural honey. Supermarket honey becomes, essentially, sugar syrup, not honey. In fact, much of the 'honey' has beet-sugar syrup added to it. A large proportion of honey in supermarkets is now imported from China, and some has been found to contain chemicals banned in Europe and the US, like chloramphenicol, a broad-spectrum antibiotic.

Our honey is harvested three times over the course of the spring and summer. Each harvest reflects the characteristics of the season, and is coloured by the flowering plants the bees are visiting at the time. Early-spring honey, which comes mostly from the fields of flowering canola, is creamy and white. Late-season honey comes from the flowering chestnut trees and becomes rich, dark and aromatic. Vines are self-pollinating so they don't need bees to produce fruit, but our fruit orchard thrives with the presence of the hives, as do the multitude of wildflowers and trees around the farm. When I think of the ecosystem here, and I think about how the grapes in the vineyard benefit from the health of all the surrounding flora and fauna, I see that the bees play a starring role in this kaleidoscope of life.

One morning, the smell of burning wakes me at dawn. I struggle up from the depths of sleep, my thoughts spinning: *There was no fire in the fireplace in our room last night. Is the house burning? Is the forest burning?* I sit up fast but sink back down in realisation. *The frost.* Of course. I get up and go to the window. The first daylight is just showing the dark forms of the vines in the South parcel.

I pull on some warm clothes and go downstairs to make coffee, and by the time I get outside it's light. The smell of smoke stings my nostrils. Walking towards the North parcel I look out on a scene from *Apocalypse Now*: the valley below is bathed in smoke, plumes rising towards the sky, fires flickering here and there around the vines on the plateau on the opposite side of the valley, like the burning aftermath of mortar strikes. In the distance, the rhythmic thumping of two helicopters hovering over the vineyards. A battle against frost is underway.

These spring frosts are a calamity for winegrowers in numerous regions of the world, but particularly in the northernmost vineyards of Europe. The further north you travel from the Mediterranean and east from the Atlantic, the more common and devastating they are. In France, the inland Loire Valley, Burgundy and Champagne tend to be in most danger. The old-timers say that this type of event used to happen maybe once a decade, but recently it has been every two or three years.

It seems paradoxical that global warming is making these events more common, but because the weather is unstable and winters are becoming milder, with bouts of unseasonal warmth at the very beginning of spring, the vine sap starts to flow and the buds begin to swell early. Once they open, even if ever so slightly, they are susceptible to frost. April truly can be the cruellest month, just when the vineyard is at the cusp between winter's last gasps and the straight run into the warm days and nights of summer.

Being at war with frost is a treacherous business. How can you beat the weather? There are essentially two approaches: heat and wind. Anti-frost 'candles' are buckets of paraffin positioned throughout the vineyard, lit when the temperature drops below zero. On average you need to put around 200–300 such 'candles' every couple of acres to generate enough heat to have any effect on the vines. Each bucket weighs around 7 kg and costs around 7 euros. Putting them in place, lighting, extinguishing them, packing them back up and storing them takes a tremendous amount of time and energy and, therefore, cost. Although less toxic ones are becoming more common, burning hundreds of these candles for hours isn't good for the local air quality, and when they are finished the metal buckets are put into landfill. Many winegrowers also put huge bales of hay at strategic points around the vineyard and burn these as the sun begins to rise, creating a smokescreen to block the sunlight – it is actually the heat of direct sunlight that burns the frosty leaves and buds.

Some farms use wind machines, which resemble huge fans on a tower about 10 metres high. These are designed to reverse temperature inversion – cold air sinks to the ground level where the vines are, hot air rises. So these fans mix the warmer air down into the colder air and increase the temperature. The helicopters hovering above vineyards are trying to do the same thing, move and mix the air to warm the vines.

When you learn about frost protection in viticulture, the very first rule is: *Choose a site to plant where vines aren't likely to get hit by frost.* Easier said than done. In France, most of the well-known vineyards have been delineated over centuries, yet climate change has accelerated rapidly in the last decade. Site is still critical, however. It is what tends to save us on our vineyard. Vines planted on slopes, especially where there tends to be a breeze, are less likely to freeze, because the heavier cold air sinks down the slope and gathers in valleys or lowlands.

Our vineyard is approximately 150 metres above sea level, on the edge of a hill that slopes gently downwards. We've never taken any steps to fight against frost, aside from a little discreet prayer to Bacchus, the Roman God of agriculture, and maybe an anti-frost dance around a campfire. Twelve years of good fortune have meant that, even during the worst frost events in the area, we have only lost a tiny amount of our harvest. But that all changed in the spring of 2021. It was the hardest frost event for over forty years, with overnight temperatures sinking to −6°C in Vouvray.

It took some time to fully assess the frost damage, but as I walked our vineyard the following morning I could see that we had been hit hard. Over the next few days, the buds turned brown and the tiny new leaves curled up and dried out. It became clear that we had lost at least 60 percent of the year's crop. If I had become complacent about frost, having had the luck to escape it numerous times when it hit the area, this was yet another lesson in humility.

Navigating the hard times is part of farming, and part of this place. I know the ups and downs are nothing new, and we are part of a much longer story of the farm. This became vividly clear one April day, quite early on in my time here, when a car pulled up into the front courtyard. A man somewhere in his mid-thirties got out, alone. I was in the vines, watching from a distance. He didn't see me as he stood in front of the house, looking up and down as if examining the façade for something hidden. He then turned and walked slowly to the flowerbed at one side of the courtyard, squatted, and reached out to touch a bright-red tulip growing among a spray of other flowers.

I came up out of the vineyard and approached the yard. Worried that I might give him a fright, I scuffed my feet on the gravel to signal I was there. He stood and turned, revealing the tracks of his tears on both cheeks. He smiled brightly and reached his hand out to shake mine. 'I'm very sorry to bother you, to turn up unannounced like this.' He spoke shyly in French. 'My memory is filled with this place.'

Hubert turned out to be one of the grandchildren of the previous owners. He had spent all his summers here growing up, and when the farm was sold he and many in his family were heartbroken. Hubert told me that he was so torn up, he could never imagine coming back. It would be too difficult knowing that he'd never be able to stroll through the vineyards again, climb his favourite tree or sit in the orchard listening to birdsong and reading. But he couldn't stop thinking and dreaming of this place, and so after a few years had passed, he summoned the strength.

Juliette and I walked the grounds with him as he told us stories of his 'endless summers' here with his siblings, cousins, aunts and uncles, parents and grandparents. All the children, ten or so, would sleep in the *dortoir,* a large open space in the attic, with mattresses laid out on the floor. The nights were filled with games and stories, laughter and tears. The days brimmed with work and play. With so many people staying at the house, meals were a major project and everyone, including the children, had to pitch in. He told us of the table set up for the kids in the enormous pantry where they would peel potatoes and carrots they had picked in the *potager.* There were days spent making jam and other conserves. Of course, there were also days spent bottling wine and labelling, all by hand, pasting them the old-fashioned way, with milk, the older children stealing mouthfuls of wine when the adults turned their backs.

As we finished touring the grounds and the vineyard, we asked him if he would like to go into the house and look around. We could see that he felt this was an intrusion too far, but his desire was bursting, so we insisted. We walked through the house together, all of us silent. I could sense the weight of this moment. He had been in a type of mourning since the family had left, and it was time to put an end to it. Occasionally he would speak softly, a few words at a time: 'This was Grandfather's study', 'Aunt Catherine always slept in this bedroom', and when we got to the attic *dortoir* there were no words, only a broad smile shining through tears.

After Hubert's visit, the word got out that the American, the new owner, saw himself more as a caretaker, and was as interested in the past of the farm as much as its future. Since then, we have met and sometimes hosted lunch for Hubert's brothers, parents and several aunts and uncles. We listened to their stories about this place, asked them innumerable questions, tasted the old and the new wine together, all with great pleasure for us, feeling each time that we were becoming more and more steeped in the ambient history. The smoke signals went even further, beyond Hubert's family, to the minor aristocrats who inhabited the farm in the first half of the twentieth century – one of the grandchildren brought his children to visit, and it was the first time they had ever seen the place their father often talked about so fondly. We were also visited by people who used to work here, including one of the *bonnes*, live-in maids, who had witnessed the last vestiges of that old world of privilege, when the gentry or bourgeoisie could afford labourers and staff.

Then one late spring day, an elderly woman, perhaps in her nineties, was driven up to us by her son. He had been in touch earlier to ask if he could come by with his mother, Jeanne, who had been on the kitchen staff in the 1930s. She started working at the house when she was fifteen. She was a treasure trove of information, telling us about how the house was configured and run at the time, what the farm produced, where all the various crops were grown and, of course, how the wine was made – not that differently to how we were doing it, as it turned out. When they arrived, Jeanne asked us to pull the chain that rings the big wrought-iron bell still firmly attached to the façade. The bell used to ring out across the vineyards and fields every day at mealtimes when she worked in the kitchen, signalling to the workers to come in for lunch. And it still worked, peeling loud and strong, echoing around the land.

After her first visit, we invited Jeanne back on a day when the horses were working. She hobbled down towards the East parcel with her cane to watch Philippe and Mascotte tilling, her eyes transfixed as she tumbled back in time in her mind. Every year, Jeanne still writes us a letter from her nursing home, wishing us well in the New Year. We respond in kind, but every time I sign off I can't help but wonder if it will be the last letter.

As spring settles, the buds on the vines begin to show the first leaves. The shoots and tendrils begin to push out, too, and on every fruiting branch we begin to see the minuscule bunches of pre-grapes. These baby clusters are the size of my thumbnail. This is the moment when we first get an idea of how bountiful the harvest will be. Because our approach is low intervention in the vineyard, the yields vary each year. This is not to say they can triple or even double from one year to the next, unless there is another frost catastrophe, but a variation of 10 to 20 per cent is not unusual.

Before I was a winegrower, I would slip into a nineteenth-century fantasy when I thought of a vineyard in spring, imagining a galaxy of flowers. If every individual fruit comes from a flower, I thought, a vineyard in bloom must be a kind of Eden. It's true that every grape comes from an individual flower, but these flowers are so small that even in an entire bunch of future grapes, you can't really see them unless you kneel before the vine and look closely. It is a Lilliputian world. A universe that goes about its business with hardly anyone noticing. From a distance of more than ten yards, you would have no idea that a couple of acres of grapes were in front of you holding millions of flowers. In the Loire Valley, the chenin blanc vines flower in the first half of June, with tiny spiderweb-like filaments, pistils and stamens huddled together in an embrace of delicate yellow and white. This embrace has been repeated for millennia, dazzling the men and women guiding nature to make the things that enrich their lives. As the vineyard comes into flower, even if I can't see anything from afar, I know there is a miracle happening in close-up.

There is ethereal and pristine beauty here, but there are also many things that can go wrong. With the shoots growing a few centimetres each day at peak-spring vigour, the vines are vibrant and full of energy. As the tiny flowers open, winegrowers hold their breath. If the shoot thinning has been completed and the first treatment of the vines against fungal disease done, then nature must simply take its course. The flowers may be self-pollinating, but the weather complicates things. 'Fruit-set' is a moment of truth, as a flower becomes the beginning of a fruit, and if it's too cold, too windy, or if it rains during flowering, fruit-set can be affected, resulting in *coulure*, where some grapes are tiny and some bunches don't develop properly. All these problems have an impact on both the quality and quantity of the harvest.

Flowering is one of the many times in a winegrower's year when they must wait and surrender to fate. Over the years, this surrender has become a mentor of sorts, teaching me patience, forcing me to develop the strength that it takes to nurture and stay calm, to become like some sort of viniculture monastic. We've been trained to intervene and control, to direct and lead. To be pragmatic and take action is a modern world mantra. From classrooms to boardrooms, we are taught to act. We *must* control ourselves and shape the world around us. It's rare that we are taught to let go, be idle and indecisive, to wait. I have certainly not mastered any of this. Far from it. But the vineyard keeps showing me that control and proactivity, intervention and engineering are not always the best approach, and can often be detrimental. When there is nothing more I can do to help the vines, and I have no choice but to hand it all over to nature, I try to do it willingly, without stress, because remarkable, unexpected things happen, strengthening both the plants and me.

When our children were young, spring conjured magic. It always seemed as if, suddenly, the vineyard became a sort of petting zoo. There were babies everywhere, furry and feathered, furtive and docile, some even within reach of a cuddle. I can still see them all: Daphné chasing a skittering line of pheasant chicks into the tall grass near the west plot; Célestine sitting on the ground in a row vines near the woods on the north side holding a young hare in her arms, petting its head lightly before it sprang off into the undergrowth; Alex cupping a tiny robin chick in the palm of his hand, almost in tears: 'Papa it fell out of its nest, you have to help me put it back!' There were sightings of fawns teetering on their stick-like legs, venturing their first steps; a string of tiny boar scampering after their mother into the woods; baby bats in the scaffolding under the old, unused water tank; an air-borne mosaic of newly hatched butterflies down by the bush of summer lilac near the entrance of the driveway. All of these became annual events in the children's lives, as exciting to them as the approach of Christmas or their birthday parties.

In spring the family energy level also goes up, like sap, and is expressed in the love and affection we share. Now the children are older, two of them already off and out into the world, Juliette and I have become eager for the arrival of the next generation, and dream of watching our grandchildren live the same seasons with us. But the truth is, we loved this spring communion with wildlife as much as the children, because it meant we became children with them. And even if they aren't here in the spring, the little kids in Juliette and me still make an appearance every year.

The counterpoint to this magic are all the herbicides, fungicides and insecticides that have been dousing vineyards since the 1950s. Organic farming largely restricts this but still allows far too much in my opinion. In the Loire Valley (and most winegrowing regions), the biggest potential biological danger to the harvest (I say 'biological' because hail and frost can be more problematic) is fungal disease: downy and powdery mildew. Most of the parasitic problems in French vineyards were actually brought from North America, and I've been the butt of numerous jokes from my French colleagues about how all their problems originally come from my homeland. *Phylloxera*, an almost microscopic insect that devastated European vineyards in the nineteenth century, came on plants that were brought on trading ships from the New World, as did many of the most prevalent fungal diseases that now affect European vines. Whereas American vine varieties have evolved an immunity to all these parasites, the *Vitis vinifera* (European wine grapes like chardonnay, cabernet sauvignon, pinot noir and chenin blanc) have not. Many in Europe have tried to grow and make wine from North American varieties, but most results are not very palatable. All grapes are not equally fit for purpose.

The presence of these fungal diseases have led to a plethora of offerings from the chemical industry. Walking the pesticide aisle of an agricultural supply depot is a bit like walking down the cereal aisle in an American supermarket: the choice is mind-boggling. If you are farming organically, however, it is blissfully simple. For powdery mildew: use sulphur. For downy mildew: use copper. These two chemical elements have been used ever since these diseases arrived on European shores. Although these substances are found abundantly in nature (copper is essential to all animal life) I still use them very sparingly as they can impact life beyond the diseases in the vineyard. Small amounts of copper can be absorbed and processed by the soil and its inhabitants, whereas too much copper can render a soil lifeless. Here on the farm, we use less than one-third of the amount permitted for organic viticulture. In order for these small doses of copper and sulphur to work, we have to adopt an extremely rigorous approach to canopy management, and not fool ourselves with overly ambitious yields.

In truth, managing diseases in an ecologically harmless way

is not that difficult. What can make it complicated is ambition. If you are a vineyard that is caught up in the endless race for bigger yields and lower prices, a race that supermarkets actively nurture and encourage, it is likely you will have to submit to the pressures of production and intervene more heavily with the chemicals. It will feel like you have no choice, in order to keep up with the market, to keep up with neighbours and the industry as a whole. Indeed, we as a society are now so removed from the production of any farm-grown drink or food that we don't realise that the prices we pay aren't cheap. It's just that someone else is paying the high price: the farmer, the community, local wildlife, rivers, and even ultimately, the consumer, who can pay the highest price of all with the loss of health, both theirs and the planet's. Standing in the supermarket, an 'inexpensive' bottle of wine seems like a good deal in the moment, but prices don't reflect the true costs.

That's why I attach the sprayer to my tractor with such a heavy heart. The three-point hitch system on the back of the tractor will lift and carry it, weighing around 400kg when filled. The transmission drive arm is attached to the power take-off, which will drive the pump, and I plug the machine's hydraulic cable into a plug on the back of the tractor, which will power the turbines that blow the spray out across the vines in a powerful stream of wet air. I run the control panel lead through a trap in the back window of the tractor and attach it to a rail inside the door, then plug it into the battery-power outlet next to the tractor's dashboard. The panel allows me to turn the spray nozzles on and off and control the flow rate of whatever liquid I am spraying.

As I rumble through the rows uneasily with my sprayer, I try to reassure myself that if grapes are to be grown and wine is to be made, I am doing it in the most respectful way possible. I also remind myself that ever since I started farming, we have made mixes of field horsetail and nettles to help in the management of fungal disease, gathering these two plants from the land, making teas and fermented blends for our spraying programme, even before similar products were available at local viticulture supply shops. Our vineyard is a living, breathing ecosystem of which the vines are only one piece of the puzzle.

Ruminating over things like this, daydreaming of how to be even more in tune with nature while driving a tractor with a 400kg sprayer on the back is not advisable. You should just concentrate on driving. The sprayer, being both heavy and lifted, significantly changes the tractor's centre of gravity. One spring a few years ago, driving down the steepest slope of the farm, I came to the end of the row and turned to go back up the next, no doubt mulling over some arcane winegrowing matter, when the world suddenly shifted into slow motion. It was as if something was tugging at the back end of the tractor, because its huge tyres were losing grip on the hillside. I realised, with horror, that the tractor was going to flip.

This is not that uncommon, even for more experienced tractor drivers than me. Unlike most cropfields, vineyards are often planted on slopes and on rocky soil, and every year in France there are tractor accidents, some of which result in serious injuries, even death, with the driver being crushed under the tractor. As the machine began to lift from its wheels on one side and started tilting downhill, the terror I felt at the first realisation gave way to a strange calm. The fall probably only lasted a second or two but seemed much longer. I had time to think multiple thoughts: 'I'm going to die. I want to be with my wife and children. Who will finish spraying? Who will keep the vineyard going?' And, most significantly, 'At least I will die in my vines.'

The impact was an explosion. I was slammed to the ground on my shoulder and the glass from the opposite door shattered and showered me in thousands of pieces. Fortunately, the glass was tempered and so the pieces showering me were small granular chunks rather than jagged shards. I got away with a few scratches and a badly bruised shoulder, nothing more. Insurance covered the damage. A friend came and finished spraying the vineyard with his equipment. Life goes on.

What has marked me most since I crawled out of the tractor and stood, stunned, looking at the overturned machine against a backdrop of rows of vines climbing up the hill, is how the only thing that mattered to me in that moment was my family and the vineyard.

When I think back to the many things I had already done by the age of thirty-eight, when I decided to become a winegrower, I am staggered at how little satisfaction I found. It's as if I saw my life as being about racking up points or ticking off boxes of places to go, things to see, taste, do, acquire, own.

Sometimes I wonder what I should have done differently. What advice would I give to my younger self? What do I tell my children? Is there any real advice to give beyond the well-worn, 'Do work that you enjoy.' Or is this far too individualistic and self-absorbed too? *Please yourself.* Is the *chemin de la vie* of each person up to them to find and follow? I still have no idea what the answer is really, though at times I have an inkling. Like the distinct sense of happiness the day I showed my young daughter a praying-mantis egg sac in the spring sunlight, attached to a post in the vines. The look on her face was all I needed. It meant that we had nurtured this place back to an equilibrium where mantis eggs could attach to posts and vines. And it meant my children could experience this with me, and were learning to appreciate this side of life, to develop their own link with nature. I couldn't ask for more.

My daughter asked me if we could take the egg sac inside and put it in a terrarium. While I am generally loathe to interfere with nature in this way, especially since praying mantises

– a great devourer of aphids and fruit flies – are an organic winegrower's best friend, I conceded. The desire in her eyes was just too much. And several weeks later I sat with her in front of the terrarium, watching dozens of mantis nymphs hatch; a whole colony of tiny, delicate mini-mantises, almost transparent, started to move before our eyes.

After pruning in winter, shoot-thinning in the spring is probably the next-most laborious vineyard task. Vines are vigorous. They want to grow far and fast, to put out as many shoots as they can. Pruning is the first stage of limiting this growth, shoot thinning is the second, and all in the name of getting the vines to focus their energy on a limited number of bunches of grapes rather than producing a plethora of shoots and leaves, many of which would not be fruit-bearing.

Pruning and shoot-thinning are similar, in that the task is both daunting and momentous: first kneeling, crouching or bending over every single vine in the vineyard, evaluating what needs to be done, bearing in mind how many bunches you think the vine should produce that year, and how you want to be shaping the vine for the next pruning season. Then you start snapping off the unwanted shoots with your fingers or a shoot-thinning knife. However, the two activities are dissimilar in two important ways. Firstly, the time pressure is much greater when you are shoot-thinning in spring, because the vines grow so fast. You need to let them grow enough before thinning to be able to see the vine in its fullness and to be able to identify the hardiest fruit-producing shoots. So you need to move quickly because, as they grow, the unwanted shoots are using energy. Also, if the shoots grow too long it's hard to put in place the trellising wires. Secondly, because shoot-thinning occurs in mid-to-late spring, there are often days when it gets quite hot. In winter, at least you can put layers on to protect yourself against the cold and the wet; days on end under a beating sun are more difficult. The best solution is to start very early and finish early, before the sun is high, but because there are so many shoots bursting into life, you often have to carry on working as the heat rages or stop and finish late into the evening.

What I love about shoot-thinning is being so close to the essence of the vine, that very moment when they are in their peak vigour. It is all energy, of earth and life. In the days it takes to get through one plot, the last shoots we break off will be twice the length of those removed at the beginning. The tendrils unfold. The leaves unfurl. The soft green trunks of each shoot seem to pulsate with nutrient-rich sap. A renaissance. The vines' energy in springtime gives me energy, too. I can feel myself gaining in strength and zeal. Working longer days as the days get longer; daylight time expands to give us ample space to finish what needs to be done. The soft green shoots fall to the ground, to return to the earth and nourish the vines. The vines remind me of seeing someone before and after they get a haircut. New and tidy, younger-looking, ready to go.

Speaking of vines as though they're sentient seems like madness to most, but when I break off a shoot I often wonder if there is a little scream of pain, sound we just can't hear because our ears aren't sensitive to that frequency. *I hold a beast, an angel and a madman in me*, Dylan Thomas wrote. I try to keep my madman at bay, but the more I live with and for plants, the more I sense him. I had never *really* noticed the natural world before. Now, the more I notice and learn, the more I see things that are fantastic and surreal: trees dancing, animals playing, insects in conferences. Why not plants crying? Is it possible to say that nature is supernatural? This may be a contradiction, but somehow it feels clear and true to me.

The past few decades have seen astounding scientific discoveries about the life of plants, most notably, forests. It's looking more and more as though forests are to plants what cities are to people. The diversity, the hustle and bustle, the endless noise and communication is all there, albeit on a level that *Homo sapiens* is not easily able to perceive. There is incontrovertible evidence that plants communicate with one another, both above and below ground. In the air they do so by releasing chemicals, volatile organic compounds, which can serve to warn their neighbours of dangers such as insect infestations. Below ground, plants' root systems are interlinked with those around them, and they 'speak' to one another using soluble compounds exchanged by roots and networks of threadlike fungi. There is now evidence that plants may also communicate using ultrasonic sounds and that certain plants recognise their kin and interact with them differently from non-relatives.

Mad I may be, but I have met even madder people on this journey. Jacques Armand Arthur Louis Puisais, for one. Oenologist, biologist, philosopher, author, teacher, and one of the best-known personalities in the French wine world in the last century. Christened *le pape des papilles*, the pope of taste buds, his service to France through food and wine was recognised by the President of the Republic, who gave him the *Légion d'Honneur* medal. Through the proverbial grapevine he had heard about our artisanal wine project back in 2009, and was intrigued. The pope was coming to visit us.

As an outsider and foreigner, I had never heard of him. So I read to find out more: *A visionary, Jacques Puisais was the first to consider that taste is not simply a question of flavour, but a global perception of the product: its texture, its smell, and a whole emotional relationship that surrounds it. With poetic sensibility, he campaigned for the 'true' taste: the famous triangle which connects a person, a product and a situation. An experience without artifice or sophistication, like 'the pleasure of eating a hard-boiled egg on the counter of the buffet at St Pierre-des-Corps station before taking the train.' In the 1970s, he created a sensory-awakening method that brought taste into primary schools: a real 'palate revolution' for thousands of children who learned to put words to dishes, and to share their emotions with their classmates. Tasting a simple apple becomes a unique moment that will anchor itself in the memory, the moment when everyone expresses themselves and constructs their own evaluation scale. This is how shared emotion is born.*

In 1976, Jacques Puisais co-founded the French Taste Institute in Tours: he brought together pharmacists, doctors, sociologists, psychologists, historians, restaurateurs, leaders of the agro-food industry ... all gathered around exploring the idea of what food is, and what it should be.

This all might sound 'very French', but the confluence of place, produce and people intrigued me. Juliette of course knew of him, and reminded me of how our children had talked about *La Semaine du Goût*, or Taste Week, at school every year throughout nursery and primary school. It is time dedicated to learning about food: tasting and describing different fruit, vegetables, spices and dishes, along with visits from local chefs describing what it means to prepare a good, healthy meal. Most importantly, they were given fresh ingredients to prepare their own meals. This, she told me, was part of Jacques Puisais' legacy to France.

Monsieur Puisais was born not too far from here, in Poitiers, in 1927. By the time he reached us on the vineyard he was into his eighties. But his age had done nothing to blunt his energy, enthusiasm and insight. 'Please, call me Jacques,' he said, holding out his hand.

We spent the day walking gently around the vineyard, Jacques peppering our conversation with wonderful anecdotes and gems of experience gleaned from a life dedicated to taste. After the tour we sat down for lunch, and as we neared the end, just before the cheese course, I poured us all a glass of our 2010 vintage: a full-bodied demi-sec with a vibrant acidity which balanced out the sweetness beautifully. We all took a sip and Jacques smiled. 'Remember what you just tasted,' he said with mischief in his eyes. 'Now, take your glasses and come with me.'

We followed him down into the cellar, where we all tasted the wine again. We then left the house and followed him into the woods, where we did the same again. He wanted to show us how the wine tasted different in each of these locations, gently pulled in one direction or another by the damp stone smell of the cellar or the earthy, mulch smell of the woods. He suggested that the darkness of the cellar and its confined atmosphere imprisoned the wine somewhat, dulling its expression, whereas tasting it outside in the spring sunshine gave it freedom. I have never done wine-tasting with clients in the cellar since then; we have always set up the tasting tables outside in the fresh air.

On the way back from the woods, in the yard, Jacques instructed us to hold our glasses up against the deep-green backdrop of the ivy growing on the façade of the house. 'Stare at the green for one full minute, through the wine. Now, close your eyes and picture that green, and taste the wine.' We did the same with the blue of the sky and the grey stone outside the winery. There were moments when he did seem a little bit mad, but he was a person attuned to the sensory experience of eating and drinking in a way that few others are. I might not have noticed all the nuances of taste influenced by the colour or spaces we were in that day, but I definitely did get some of them. Without doubt, Jacques sparked a desire in us to learn how to get more of those experiences, which is something Juliette and I continue to strive for – in the kitchen, dining room, vineyard, cellar, farmyard, and sometimes among the vines and woods.

Even if I occasionally wonder whether breaking off shoots might not be the most pleasant experience for the vines, my empathy only goes so far. I want to make the best wine I possibly can, and shoot-thinning is crucial to this. Even some industrial winemakers, in an effort to boost quality, go through their vineyards with shoot-thinning machines. There are different types of machines, but the principle is the same. Mounted to a tractor, the machine features two spinning heads of spaghetti-like whips that whack off shoots on either side of the row. They are not designed for shoot-thinning but de-suckering, meaning that they can only get at the sucker shoots low on the vine, and cannot thin the shoots higher up nearer the fruiting branches. There's a big difference between this and a person spending time in front of every vine, making decisions about which shoots to take off by hand.

Depending on the season and the vigour of a given vine, we might remove as many as half of the shoots. You can imagine the effect this has on the number of nutrients delivered by the plant to the remaining shoots, not to mention the additional energy provided by the sunlight with fewer branches blocking them. In fact, shoot-thinning by hand may be the most important canopy-management activity in the vineyard, and I can't do it alone because there isn't much time to get the job done.

In all French winegrowing regions there's a sub-economy of seasonal workers, both locals and people from different regions and other countries. The local workforce tends to be groups of friends, often young people, who, along with the energy and enthusiasm of youth, have the sort of knowledge and wisdom that can't be taught. They are generally not from privileged families, and while some work in vineyards out of absolute necessity, others do it by choice, having rejected mainstream society and culture, living instead a nomadic, hand-to-mouth life in their vans or campers. Running through all of them is an enviable camaraderie, a respect for hard work, and a sense that life is not, and never was, a long tranquil river. An appreciation of the precariousness of employment and income makes them older than their years, having had to make choices and reflect on things that their more privileged compatriots don't have to.

Even though the minimum wage in France is relatively high, it's getting more and more difficult to find local people to do agricultural work – wages aren't great and the work is hard. Over the last few years, the gaps this dwindling workforce have left in the countryside have been filled by teams of itinerant agricultural workers from the less prosperous countries in the EU. At our farm, as in the surrounding area, these teams generally come from Bulgaria. In contrast to the local workforce, the people in these groups tend to be older and are often educated and highly skilled. I suppose these are the people left behind by European integration and globalisation – as Bulgaria develops, the younger generation is becoming more educated, more 'European', and like the same generation here in France, they have a higher expectation of what their career should look like and how much they expect to earn. It's the older generation that find themselves out of work or underemployed.

For as long as the income disparities remain between rich and less rich countries in Europe, it makes it worthwhile for these groups to travel from Bulgaria and other countries to work. Usually, they stay for several months, working intense, long hours, before going home for a time and coming back, rotating in and out of the various regions and seasonal activities going on in French farms.

I have met ex-university professors and engineers, who all tell me there simply isn't enough work for them in Bulgaria. Speaking to them, it has dawned on me that there's a widespread notion in the 'rich' northwestern European countries that somehow life owes us something. It owes us an excellent education, healthcare, job security, material wellbeing, a social-security safety net, a pension. We see the catastrophes and misery, the wars and despots making life difficult or even miserable for most of the human population, yet somehow it doesn't touch us. The cost of living may rise and wages don't go so far anymore, but we are generally safer and healthier than most, while our children have many more opportunities. What would be considered by many in the world to be luxuries, have become basic rights and liberties to us. The Bulgarian workers in our vineyard still live in a not-too-distant past, when most of Europe was not so privileged, when social justice hardly existed, and nations were still war-torn and impoverished.

One day, a Bulgarian team showed up to help with shoot-thinning. Five people emerged from a twenty-year-old Peugeot with Bulgarian plates. Among them were two elderly women who could have stepped out of a Tolstoy novel, both of them small and frail with weathered, wrinkled faces, wearing headscarves and baggy homemade peasant dresses with aprons that would have been familiar to serfs of pre-revolution Russia. We could not communicate. Not one bit. I was somewhat embarrassed, feeling both upset at the injustice that forced these old women to continue doing backbreaking physical work and, at the same time, a bit worried they'd be slower so I'd end up paying more for shoot-thinning than I had budgeted. It turned out, perhaps unsurprisingly, that both of them were strong, fast and extremely efficient. They had been doing work like this all their lives. It was child's play to them. Though we had no language in common, when we took a coffee break together we exchanged smiles and I pondered the stories carved into their walnut cheeks.

My friend Damien's thirtieth-birthday party was like a spring festival – Touraine's equivalent of Northern India's Holi, celebrating love and the beginning of spring, without the powder-throwing and face-painting. But the music, singing, dancing and feasting were all there.

We arrived a bit late to his winery, across the river from us on the outskirts of Amboise. We wandered into the long, broad field next to a plot of his vines where the festivities were already under way. Under a tent, a band was playing swing tunes. Damien, apart from being a talented winegrower, is also an accomplished clarinettist, playing with the local orchestra in his spare time, if there ever was any. A friend of his, a local brewer, had set up a beer truck at one end of the field, and alongside this there was a row of wine barrels, spigots twisted fervently by partygoers. Off to one end of the field, there were two huge spits smoking away, several suckling pigs rotating above the flames. Two long tables were overflowing with an array of salads, cheese and small mountains of bread. Deckchairs and bales of hay were strewn across the field to sit or lie on. Some people were already dancing in front of the band. It was early afternoon, and we could already tell that this was a party that would go on all day and night.

There were multiple generations: a group of very old-timers sitting in a cluster, taking in the festivities, watching children of all ages running about. Juliette and I knew many of the people already, most from winegrower families, the web that holds these communities together. But Damien's reach went well beyond that. And that day, beyond the people we already knew, Juliette and I met a baker, hairdresser, doctor, furniture-maker, teacher, musician, local politician, and many others who never mentioned what they did – if it weren't for the American in me, I might not have known what any of these people 'did'. Juliette noticed and mentioned it to me afterwards, how quickly, how early in the conversation I went for the question, *What do you do?* Over the years, I have learned that this tendency to put a label on someone is much less pronounced here. The conversation is the thing, not the position of the individual. You get there eventually, but in due course, not as the main subject.

As the party went on and the sun began to cast longer

shadows across the field, I pondered what it was that held all these people together. Wine? It would be difficult to find someone at the party who didn't like it and drank some most days. Wine is as much a part of living here as the sight of the Loire River making its way to the sea. Yet I think the real bond is living from the land, and in this place. It is about living in and around Amboise, this beautiful Renaissance town, with its castle standing firm on the hill, overlooking the entire river valley, where generations of the kings of France lived, from Charles VIII to François I. It is knowing that, if you raised your eyes while walking the cobblestoned street to the bakery or the butcher, you would see the flamboyant gothic chapel of Saint Hubert, built in 1493, its grand, lead-covered spire adorned with golden stag horns, reaching towards the sky. And perhaps you would give a thought to Leonardo da Vinci, who died in Amboise in 1519, while living under the patronage of François I, and whose remains are in a tomb in that very chapel. The chapel itself, named after the patron saint of hunting, might bring to your mind the forest of Amboise, a huge sweep of ancient kings' hunting grounds, which to this day teems with wildlife. Maybe you would then picture bands of Vikings, hiding out in that forest before they sacked the town in 853AD, or imagine the advance of Alaric II, king of the Visigoth invaders and his army, marching through those woods in 504AD. As you live and grow up here, a deep sense of history gets under your skin.

Propped on a bale of hay watching the party unfold, I remembered visiting places like this in Europe when I was younger, before I moved here. It felt, at the time, like visiting a series of postcards, each more strange, impressive and majestic than the previous. 'Do people really live here?' I would ask myself. 'What do they *do*?' Now, at least, I have some of the answers.

The language of viticulture is European, once Latin, now mostly French and Italian. And as I install the discs on my tractor, I wonder vaguely how in English to describe what I'm about to do, realising that I have no idea. The French call it *chaussage des vignes*, which, if you translate it literally, would be 'shoeing the vines', as in shoeing a horse. I take out my mobile phone, the supercomputer in my pocket, and look it up. *Ridging the vines*, apparently.

Essentially, I will be reversing what I did in very early spring with the retracting mechanical hoe. The discs behind the tractor are tilted outwards at around 30 degrees, and as the tractor (a horse can do this too) pulls the discs, they dig into the earth about 30 centimetres from the base of the rows of vines and, turning, push the soil that the hoeing had removed from around the trunks back up onto them. This covers up any new undergrowth around the vines: weeding by burying. It also regulates competition for nutrients between the vines and the cover growth. This push-and-pull cycle of scraping earth away from the vines, and then shoving it back onto them again, is the basis of organic control of cover crops. At this same moment, just down the road and in most vineyards all over the world, tractors will be going through the vines not with discs but with sprayers filled with glyphosate, spraying the earth around the vines to kill unwanted plants.

Chaussage is both gratifying and useful. From a purely viticultural perspective, it is essential to keeping the vines from becoming overgrown. But I find it satisfying, too. The long straight lines, discs churning, moving down the undeviating narrow paths – the symmetry is pleasing to the eye and mind. After a day of this, I pull myself down from the tractor and look at the vines, walking around the parcels. The uniform mound of earth rolling down every row, cut almost razor-straight with the steel discs gives me a calming feeling. Driving all day down the rows is also somehow liberating. There is no choice, you can't turn right or left, you can't deviate from the row, straight on is all, unflinching. Alan Lightman, an American scientist, wrote: 'Symmetry represents order, and we crave order in this strange universe we find ourselves in ... [It] helps us make sense of the world around us.'

The birds like *chaussage* as well. As the discs cut and overturn the earth, a feast is laid out for them: worms, grubs and insects brought suddenly up from their subterranean world, exposed to the elements and predators. Crows and gulls are the Goliaths here, blue tits and robins the Davids, working their wiles to snatch their share while giving their larger rivals a wide berth. Trundling along in my tractor, with this trail of birds behind me, makes me feel like a benign pied piper, not wading into the Weser River but through an earthy cornucopia.

Last night, I had a dream of wires. Endless kilometres of wires. Electricity wires stretching on their massive steel perches across countries, undersea wires wrapped in plastic and rubber crossing oceans, fibre optic strands that could reach across the universe. I woke and opened my eyes to a room that looked like a silver halide photograph: black and white and every shade of grey, etching every surface and almost glowing in crystal moonlight.

The curtains were all fully open in the room, the sky was completely clear and, as I turned to the window nearest the bed, I saw the massive, glowing disc of a full Flower Moon, the full May moon. Over the years on the vineyard, I have become familiar with the lunar cycles, from every new and quarter moon, waxing and waning crescent and gibbous moons and, of course, the spectrum of full moons across the year. Every month's full moon has a name, all wonderfully evocative and pertinent. In January, Wolf Moon. February, Snow Moon. June, Strawberry Moon. September, Corn Moon. December is the Cold Moon. I know them all now, and observe each one every year with awe and wonder.

In my previous life I never really took much notice of the night sky. Cities don't lend themselves to it. I also used to need complete darkness to sleep, curtains always drawn tight against the outside world. Now I never draw the curtains. I live with the outside world. The nights, clear and moonlit or dark and disturbed, and all of the morning light – the sunrise, dawn, daybreak, first light, *l'aube, l'aurore.* On the longest summer days, when it doesn't get dark until after 10pm, if I'm tired I might even be in bed as the last light fades, rubbing away the shadows in my room. When I wake up from a dream or a nightmare, the room awaiting me is always different, rarely completely dark.

The dream of wires was somewhere between dream and nightmare. There was definitely anxiety, subdued by the Flower Moon. I know exactly where the wire in my dream has come from: trellising wires are an essential part of a vineyard. Different regions in France may have different types of trellising depending on how the vines are pruned and trained, but no matter what the system of trellising is, there are always kilometres of wires to manage. In winter, after pruning, the wires have to be lifted from the ground and attached to the top of the posts, to allow for tilling without breaking wires. Now, as spring marches on and thinning vine shoots is underway, the wires need to be put back down onto the ground. The wires are raised and lowered manually. At the end of spring, when the shoots have grown out and are starting to grow into the rows, the wires on both sides of each row of vines are lifted and clipped together, bringing all the shoots into an upright position. If the vines were left to their own devices, the shoots would grow into long branches, invading the vineyard, making it impossible to get through with a tractor or horse to till, mow, trim or spray, and making harvest a very messy business indeed.

So, in late spring, I tend to (and dream of) the wires, all of them made from galvanised steel to keep them from rusting for as long as possible – which is not to say they don't eventually rust and break and need to be replaced. Contrary to what one might think, steel stretches, and after a season of being taught, they become a little slack. Also, the wooden posts at the ends of each row to which they are attached will give a little. This all means that I'm obliged to spend many days in the spring vineyard unwinding every wire on every row at one end-post, pulling each tight again, and winding it back around the post. There are over 300 rows in our vineyard and each row has two wires. Unwinding, pulling tight, and rewinding over 600 steel trellis wires is another one of those seemingly insurmountable vineyard tasks. The toll it takes on my hands is not inconsiderable, even with gloves. Some nights, I dream of having solid-leather hands. Sisyphus was condemned to rolling his boulder for cheating death twice. With all my boulder-rolling on the vineyard, I like to think that I might get at least one chance to cheat that dark rascal, get him to give me one or two more harvests before I go the way of rusted-out trellis wires.

If Greek mythology provides some insight into vineyard life, Biblical mythology is pretty good, too, with its forty days and forty nights of rain, plague and pestilence. The book of Exodus should probably be required reading for student winegrowers, to prepare them for the unpredictability of this growing life. *The Lord sent thunder and hail, and lightning flashed down to the ground.*

Like frosts, hailstorms are not uncommon in European grape-growing regions. Generally, they come in the spring, bearing ice stones sometimes the size of marbles, tearing holes in newly formed vineleaves, ripping them to shreds, breaking the soft shoots to which they cling and destroying clusters of flowering inflorescence. It is violent in the extreme. Hailstorms are usually very localised, so there is absolutely no way of preparing for the sudden raid. When hail is forecast in our region, all the winemakers brace themselves for a round of Russian roulette. Not everyone will suffer, but someone almost certainly will.

In 2021, hail ripped through a nascent harvest a few kilometres away from us. In 2013, one of the worst spring hailstorms hit the Vouvray region, destroying a large percentage of the crop – plots of vines 50 metres from ours were almost completely decimated, while we had zero damage. But after the lessons of frost, I am no longer complacent. I know, if I live long enough, I will almost certainly see at least part of my crop destroyed by a hailstorm. Still, until the water of the Loire River *turns to blood, and the fish die and the river stinks*, I will persist and look forward to every spring with excitement, trepidation, joy and a *soupçon* of fear.

L'ETÉ
SUMMER

To the attentive eye, each moment of the year has its own beauty, and in the same field, it beholds, every hour, a picture which was never seen before, and which shall never be seen again.

RAPLH WALDO EMERSON

By the beginning of summer my body is covered in insect bites and stings. All that time in the spring vineyard, coinciding with endless insect hatches, has left me riddled: midges, horseflies, mosquitos, spiders, ants and the many other stinging and biting species unknown to me. There are other challenges to the body that lie ahead, but the dry heat of summer does bring some relief. If the trellis wires are up, the soil tilled under the vines, the rows mowed, flowering and pollination successful, with each bunch now formed into small clusters of tiny, pea-sized grapes, the hectic activity of spring does give way to more deliberate, slower-paced work (and fewer galvanised-wire dreams).

Below ground in the cellar, fermentation has run its course and last year's wine is almost ready for bottling. Juliette and I spend time in the musty cool air tasting each barrel, making notes of our impressions, each year developing a little more understanding of the differences between the plots of land where the grapes grow, and the effect that different seasonal elements – frost, rain, temperature, sunlight – have on each vintage. We won't bottle our wine until the end of summer, but we need to plan the blends we'll make. We also order the bottles and corks, check the hoses, test the pumps, and calculate how many and which of the larger vats we will use for blending just before bottling. In between this preparation, it's a good time to update the people who sell and import our wine, giving them a sense of the profile of the vintage and how many bottles will be available.

Tasting the finished wine in the barrels is one of the supremely rewarding aspects of growing wine. The transformation comes full circle. From the soil and the heavens to the vines and their fruit, from the freshly pressed juice to the fermenting grape must, to the wine itself, first tasted at home with Juliette. Although the joy and celebration of the new vintage will come later, when we open the first bottles with friends and family around a good meal, those first tastes in early summer give us both a visceral satisfaction and knowledge that we've played our part in this mysterious process.

The increasing hours of sunlight imbue some days with a sense that there is plenty of time to get things done, a feeling of the endless summer that I find creates spaces for the mind to wander. Memories resurface, too, such as my childhood love for climbing trees. The higher the better. Had my parents seen some of the heights I reached, I'm sure they would have been terrified.

This memory kept coming back last summer, so much so that I started eyeing up some of the tall trees on the farm – the sequoia, the oaks and conifers. The day will come when I won't be able to climb again, or probably shouldn't because of my responsibility to Juliette and the children.

I waited for a bright, windless morning, and when one came along I decided it was time. While the sequoia is probably the tallest tree, there is a spruce that reaches slightly higher because it was planted nearer the highest point on the farm. It must be almost 30 metres tall and, in contrast to the sturdiness of the sequoia, the spruce's trunk is thinner and its branches lankier, so, when the wind is at its most voracious, the spruce sways with wild abandon.

I knew this was not one of the wisest decisions, certainly not a very sensible one, but I have made it a rule to accept certain moments – once I have weighed things up, measured the pros and the cons – to throw myself in. So I put on gloves, tight-fitting clothing to avoid snags, set out with an extendable ladder that could reach the first branches, and started to climb.

Once I was into the thick of the foliage and branches, it felt as though the spruce had been designed to be climbed. The branches were solid and free of foliage near the trunk, rising evenly spaced towards the summit like a three-dimensional ladder. I still took my time, pulling myself up slowly, branch after branch, stopping occasionally to enjoy the smell of pine resin, woody and sweet, thinking about the sense of comfort and protection that the squirrels, birds and other tree-dwellers must feel, hidden and protected by the lush coniferous foliage so far from the threats of the ground.

When I arrived at the peak, the sensation was different. I sat on a branch and clung to the trunk, which had narrowed substantially as I got higher. A day that was windless lower down was gently breezy at this height. I felt exposed but exhilarated. Sitting atop the tallest tree on one of the highest hills around was breathtaking. The valleys rolled out in all directions, the vineyards no longer the deep-rooted, earthy places that held me to them when I was working. From here, their distant geometry was all lines and angles, patterns and human design. The house was a long sweep of grey-slate roof tiles punctuated by brick chimneys. The orchard and the *potager*, the fields of grass and walkways all seemed new from my perch. The light fell on them all differently, casting unfamiliar shadows. But these, too, were the places where I lived and worked.

As I looked out towards the Loire River, a silver expanse in the sun, I thought of the old life. My colleagues and friends from that time would be on holiday with their families about now, sailing in the Mediterranean or savouring luxury in some seaside resort. Sure, this is all good fun if you can get it, but as a winegrower, I wouldn't be doing those things again. And yet, sitting at the summit of the spruce tree, looking out over the vineyard, I may not have been the happiest man alive, but I felt like I was pretty close.

Held comfortably snug between the trellising wires, the vines' branches reach upwards to the sky. By early summer they have grown long and high, and new branches are thrusting out sideways from below the wires. When a breeze picks up, the vineyard whispers like wheatfields, with the tender, flexible branches rolling and swaying to the rhythm of the wind. This is the time for *rognage*, best translated as 'trimming'. We don't use a *rogneur* (a machine attached to the tractor with spinning blades that lop off the tops and sides of the vine growth), we do it by hand, so it is more gentle, more like trimming a hedge. I take up my shears and set off on a task that will last several days, and be repeated twice over the summer. The clack-clack-clack of the shears sent out across the vineyard marks early July.

Trimming opens the vineyard up, reducing the shadow cast across the rows and making it easier for any sprayed treatments to reach the grape bunches. Once again, the vigorous vines, if left untrimmed, would revert to their jungle impulse to keep reaching and growing, making it difficult to get into the rows with a tractor and even hindering a person harvesting by hand. Trimming vines requires much less force than trimming hedges. Because the vine branches are so young and tender, the shears slice through them like butter, which is some compensation at least for the time it takes. If all the rows in 10 acres of vineyard were lined up as one long hedgerow, it would extend for nearly 7 kilometres. Walking this far is one thing. Walking while trimming the tops and sides of the vines with shears is quite another.

One day I might sit down and try to calculate how many hundreds, perhaps thousands, of kilometres I've covered walking through these vines, tending to them over the years.

I never find this kind of travel tedious though. The trimming journey is always punctuated by sightings of roe deer and hare, and birds galore. The clouds will put on a show, too, and the branches sway in the breeze, making me feel like I'm in the company of hundreds of chattering, bustling beings. Sometimes the rain will fall and suddenly cool my body. Hail may be my winter foe, but its cousin, summer rain, can be a true friend.

It was while trimming once that I noticed a patch of colour out of place in the green of the cover crop, perfectly still in my peripheral vision. As I moved slowly towards it, the brown shape formed into a newborn roe deer, curled up under a lush old vine, partially hidden by the grasses growing around it. It couldn't have been more than four or five days old. I knew that the mother would be back because they leave the fawns hidden in the grass while they forage, coming back several times a day to suckle their young. The fawn wasn't much larger than a rugby ball, rusty brown with darker spots. I knelt slowly. It didn't seem at all frightened, and just looked up at me with dark eyes, not yet aware or attuned to the dangers lurking in its future. Although all alone, fawns at this age don't give off any scent, depriving predators of their primary tools: smell. I sat with her just for a few minutes, not wanting to disturb the mother should she come back, but so drawn to the calm and the beauty that I couldn't resist staying longer than I should.

In these long days of summer, I'm reminded distinctly of the bundle of often contradictory emotions felt in those early years when we first moved here. It didn't take long, perhaps just a few months into our new life, before I began to panic. Although there was still a tenancy agreement for the vineyard itself, which meant we had several years before taking up the responsibility of caring for the vines ourselves, the rest of the farm could not wait. The orchards and vegetable garden, the trees and shrubs, the flowerbeds and lawns and the meadows. It felt as if the land could not wait.

Sure, I could cut grass. But could I tend to a wildly overgrown vegetable garden? Care for fruit trees? Prune lilac and rose bushes? A hedgerow invaded by the growth of thorns, spilling out into the road and vineyard? What was that ivy doing climbing onto the roof, under the tiles, and into the guttering? It seemed to me that if I didn't get a grip on things, the entire place would soon be overgrown, turning into something resembling Angkor Wat, with thick roots winding their way through the windows of the house, branches of trees pushing through the roof of the barn.

I started to haunt the alleys of hardware and gardening stores hoping that I would somehow find the answers there. On one such visit I was asking for help from a store assistant, something about a tool for cutting back high branches in the orchard. The assistant led me over to the aisle with all the pruning equipment and left me there, befuddled, before an array of ladders, perches and secateurs. A man came up beside me. He was short, bald and a bit stocky, with friendly light-blue eyes.

'Excuse me,' he said. 'I overheard your conversation. You know, you shouldn't be pruning your fruit trees yet.' It hadn't occurred to me that there was a time of year for pruning fruit trees, but I hoped that my ignorance would be forgiven if I struck up a conversation with him.

There was something about the stranger, in his look and quiet voice that made me trust him immediately. It turned out that he had recently retired from his job with the SNCF (the French national railway company), where he had supervised teams that renovated railway carriages. He now found himself at a loose end. He loved gardening, he told me, and had been thinking about looking for some part-time work as a gardener to fill some of his free days. He was from a farming family but was the first to graduate from secondary school, which took him onto an apprenticeship as a welder. And although he had risen quickly up the ladder to his supervisory role, the farm had never left him. His family still had a mixture of livestock, and grew wheat, corn and oats. He introduced himself as Monsieur Boisnon, and for the next eighteen years he would come to our farm religiously, one day a week, through all of the seasons, regardless of what the weather was doing. The 10 acres here that were not vineyards became Monsieur Boisnon's domain.

His first name was Elian but his old-school French formality meant that I never called him that, and he never once called me Peter. He was Monsieur Boisnon, I was Monsieur Hahn. In the French custom, we shook hands every time he arrived and departed. But, oddly enough, this formality did not create a distance between us; it was about respect more than anything. We didn't need first names to understand what linked us so closely to one another.

In the early days, before I took back the vines, Messieurs Boisnon and Hahn worked together, initially reclaiming some patches of the land that were overgrown and wild. We spent long hours cutting and pulling brambles and other thorns out of hedges and shrubs, weeding, trimming tree branches. It was then that I realised that my longing to 'live in and with nature' would not be quite as I had imagined.

After one particularly difficult (and bloody) afternoon with the thorns, I decided to look up the word 'bucolic', which had been haunting me in the brambles. The synonyms were *rural, rustic, agricultural, non-urban*. There was nothing about *peaceful, calming, uplifting* or *serene*. Over time, working with Monsieur Boisnon, I learned that this new life meant, to a large extent, finding a balance with nature, which in turn meant a lot of hard work. Monsieur Boisnon was never discouraged by what seemed to me impossible projects. 'Little by little,' he would say. 'We'll get there.' His patience and perseverance were infectious, and I started to accept the tasks before me with a calmness that has served me well ever since, in the vines and on the rest of the farm.

As time went on and we managed to tame what we wanted taming, he started focusing on what he called 'hobbies': growing fruit, vegetables and flowers. He cut back all the dead wood in the orchard and planted new apple, pear and cherry trees where old trees had died. He showed me how to prune fruit trees which, like vines, need to be pruned every year if you want good fruit. He taught me to thin out the apples, peaches and pears in years where the fruit was over-abundant, reminding me that otherwise they would stay too small or not ripen. He had me help graft branches from some of the older apple trees onto young, hearty rootstock, winding string tight around the graft union and covering it in wax. The vegetable garden became his place of predilection. He would conjure up beans, peas, courgettes, cucumbers, squash and the sweetest tomatoes I had ever eaten. He set up a compost heap and showed me how to tend to it, using it to nourish his beloved vegetables. And next to them, he planted raspberries and strawberries, rosemary, thyme, bay laurel and parsley.

When the tenancy was up on the vines and I started spending more and more time in the vineyard, he and I were not together so much. Occasionally, we would still walk through the vegetable garden and orchard, each pulling an apple or pear from a tree and savouring the ripeness together. He was proud of his work and wanted to share it with me. In the later years, he began to slow down somewhat and couldn't do some of the more strenuous work he used to. He would say to me things like, 'Remember what I said about turning the compost,' or 'Let me show you again how to prune the suckers from the tomato plants so you don't forget. I'm not going to be here forever you know.'

One day, he took me into the tool room in the garage where he would change in and out of his work clothes and showed me a notebook he kept there. It was divided into the four seasons, and he had made pages of notes over the years, describing what he had done on any given day in each season. He wanted me to know that this book was here and that if he no longer was, I would know what needed to be done to take care of his places here, places that he had nurtured and was so proud of.

Although Monsieur Boisnon isn't here anymore, I still see him almost every day, outside in the greenery, in the orchard and the vegetable garden, in the hedgerows and meadow, in all those places he shaped and cared for.

Mickael the Bear (his family name is *Lours*, pronounced exactly like 'bear' in French) is somebody else who has helped me for years. We first met when I had succumbed to the flu and lost ten days pruning one winter. Mickael was recommended by a fellow winegrower and came to my rescue, working with steady determination, not speaking much, completely focused on the task at hand. At the time, he seemed to me to be in a state of hibernation, with quiet, slow movements and careful, unhurried speech. I came to know Mickael well over the years and realised that the bear in him lived with the seasons, too, and in spring and summer he exploded with strength and energy, an indefatigable worker, a mountain of willing, intelligent muscle. When he helped at harvest, picking grapes, or later on at the press, his lust for life was infectious.

Mickael's story was not uncommon among vineyard workers, in that he came from a broken home that didn't have much money. Desperate for a way out, for independence, he left school early to start working, paying his way in life. But his curiosity and vibrant intellect did not rest, and he became a fascinating and impressive autodidact. To start with, he is passionate about the natural world. Through copious reading and observation, and possessing an excellent memory, he turned himself into a walking encyclopaedia of natural history.

He is constantly showing me insects and small animals in the vineyard, explaining their life cycles, nutritional needs, unique forms and markings. But underneath all this knowledge, what struck me most was the kindness in his big blue eyes. He loves these creatures. He is truly fascinated by and in awe of them. The spiders, wasps, praying mantises, dragonflies, ants, butterflies: he knows them all. He will turn a leaf over and show me the dozens of tiny yellow eggs clinging there, and explain that they are ladybird eggs which will be hatching soon, and that not all the eggs are fertile so that when they hatch the new larvae will eat the non-fertile eggs for nourishment. 'Did you know that ladybirds hibernate in the winter, like bears!' he says with a wide grin.

Mickael is also well-versed in the vineyard flora. He points to a lacy white flower in the rows and bends to pull it up. '*Daucus carota*,' he explains, again his winning smile. He rubs the root and holds it out for me to smell – it's exactly like a carrot. 'Wild carrot,' he explains.

Mickael played a role in helping me understand the balance I was seeking in the vineyard, between what was growing and what was living there, what was indigenous and what was passing through, and how the interplay of all of these could restore an equilibrium that had been lost through 'modern' farming.

The bear didn't reveal things about himself willingly. It took time to get to know him. For a long time he had lived in his van, and the only reason he had recently started renting a small apartment was because he wanted to be able to use a computer. To my astonishment, he had taught himself how to code and had a side activity helping people build websites. He also dabbled in bitcoin. Mickael the Bear, itinerant vineyard worker with big blue eyes and permanently dishevelled hair, is still one of the most interesting people I have ever encountered.

On a hot June afternoon I was in the West plot, toiling with frustration. The trellis wires were in a terrible shape, rusting and fragile, the worst on the vineyard. When I was raising the wires and clipping them into place earlier in the season, some of them broke. I was out gathering in the old wires and replacing them with new ones. Rolling up rusty, broken wires that are 100 metres long is an unwieldy task. You reel them in, like a cowboy rolls a rope, rotating the wire in circles around the hand and elbow. If you aren't careful the wire will cut into your arm, or snap again, the pointed end scratching your skin. Often, the wire snags on a vine and breaks into another segment as you pull it free. I was sweating profusely in the late afternoon sun, bleeding from a cut on my right forearm, and swatting at the horseflies. I didn't hear the car pull up; my old friend Vincent appeared before me with a bottle of water.

'Rest a little,' he said, handing me the bottle.

While I sat down and took a swig of water, he walked down the row of vines, quiet and contemplative, a man of few words. Occasionally, he knelt before a vine looking at the newly formed bunches of grapes, or plucked a leaf and turned it over. He took a handful of soil, smelt the earth, and let it sift through his fingers. I knew he could see my frustration, feel the despair I was feeling, and maybe even read my mind and hear the voice inside my head. *What the hell are you doing here? Why did you leave all that comfort behind?*

Vincent crouched next to me and we sat for a few minutes like that, in silence, shaded partially by the tall old vines above us. Then he stood up, looked out across the vineyard and said, '*Continue Peter. Tu sais, c'est très bien ce que tu fais.*' 'Keep going Peter. You know, what you are doing is very good.' These words from this man meant more to me than any accolades I had received in my previous life, and were just what I needed to keep going through the heat to get the job done.

The hard work, the blood and the sweat always seem worth it when I'm leaf-thinning on a summer's morning, hands and forearms covered in dew, vine leaves scattered about me on the ground and lightly glowing in the sun's first light.

Plucking the leaves that surround the developing bunches of grapes, delving into the interior of the vine's canopy, is precise and rhythmic. Opening up the foliage around the bunches gives the young grapes more light and air. It allows them to dry more quickly when it rains or when they are covered in morning dew, which in turn reduces the incidence of fungal disease on the bunches. Also, if the grapes are going to be harvested manually, they will be much easier for the pickers to see if the leaves around each bunch have been removed, making the harvest faster, with fewer unseen bunches left behind.

Sometimes, a vine leaf isn't just a vine leaf. Ampelography is the somewhat obscure sub-field of botany concerned with the identification and classification of grapevines. Traditionally, this has been done by comparing the shape and colour of vine leaves. Every variety has a different leaf size and shape. Merlot, for example, has very large leaves. Gewürztraminer has small leaves. In addition to the blade size, the shape of the teeth on the outer edge, the arrangement of the five lobes, and the angle and length of the veins, all manifest differently depending on the grape.

In any case, the leaf on all varieties is the workhouse, the engine room of the vine. Photosynthesis provides the energy for the vine and the 'food' for the grapes to grow, and it's the top of the leaf that absorbs sunlight for this. Less well known is that the underside of the leaf contains thousands of microscopic pores called stomata, which open in the day and close at night. The plant essentially breathes through them, taking in the carbon dioxide needed for photosynthesis and 'breathing out' the oxygen that is so essential to us. The vine also 'sweats' through the stomata, serving the same function as it does for humans, cooling the plant down when heat levels rise. This is why the air around a tree on a hot summer day is cooler than the air in an open field, and why walking in a forest feels so much fresher than walking across a field or down a street on the same day. It isn't just the shade of the plants' leaves that reduces the temperature. It is also transpiration, the process that cools the air around the plant.

As I pluck off leaves, each emitting a little 'pop', and let them fall to the ground where they will return their energy to the soil, I consider the cultural significance of the vine leaf – the symbolism of plenty, freedom and rebirth. In mythology, the vine leaf is associated with Dionysus and his vigour and strength. Vines can also be found engraved in many old Christian churches, symbols of victory over death and the joy of heaven. The vine leaf often replaced the fig leaf in art, to cover genitalia and enable symbolic modesty. You can eat grape leaves, too. They are delicious, low in calories, high in fibre, full of vitamins A and K, and antioxidant. Spiritual and physical sustenance, all in one. Perhaps the vine leaf is best known in the Greek dish, *dolmathes*, in which the leaves are wrapped around meat, rice and herbs, a dish served hot or cold with a lemon-based sauce. Less well-known delights are vine-leaf pesto, salsa verde and deep-fried crispy vine-leaf tempura.

Grape leaves have also been used medicinally throughout history. They have been popularly used to stop bleeding, and to relieve pain, inflammation and diarrhoea. I haven't personally verified any of these medicinal uses, but as I moved, leaf-thinning through the vineyard, dropping thousands of these delicate, shapely blades on the ground, I rubbed a few leaves against my forearms, cooling the insect bites on my skin.

As the morning sun was getting more intense, Juliette came out to join me in the row. She knows that I'm too stubborn to stop before reaching the end of a plot. 'Let's get this done,' she said, starting to pluck leaves. 'I've made a delicious batch of sun tea. Just picture a big glass with ice and mint.'

'Beautiful,' I replied. 'I've been wishing for rain. Delicious, cooling rain.'

'You really are demanding, aren't you!' she laughed as she started jumping, bouncing and skipping down the row. 'This is my rain dance! It never fails!' I got up and bounced and twirled after her.

We weren't alone, or crazy. Humans since time immemorial have been trying to control the water in the skies. In ancient Rome there was a ceremony called the *aquaelicium*, 'the calling of the waters', performed in times of drought, which involved a sacred stone, a *lapis manalis*, imbued with supernatural powers. A thousand years before that the Wu shamans in China would perform the *Yu*, a ceremony in which a sorcerer would perform an exhausting dance inside a ring of fire, until he began to sweat profusely, the drops shaking off his baking body to the earth below, symbolising the desired rain. The most familiar ceremonies to us are probably the Native American rain dances, common to many of these ancient tribes, their origins deep in prehistoric time. The colourful magnificence of the costumes and the rhythmic precision of the dancing remain part of their culture to this day.

What strange contortions we winegrowers go through to control the skies! A few years ago in Vouvray, when the threatening, anvil-shaped cumulonimbus clouds began to gather in late spring, the hail rockets started firing – the Vouvray wine *syndicat* had decided to invest in a system of these rocket launchers, which look disturbingly like those you see on television news, blasting death towards soldiers and civilians. The rockets stab into the clouds and explode, disseminating a spray of silver iodide droplets, which hinder the formation of hailstones. The effectiveness of this technique, however, has been shown to be limited.

As the climate changes, water has become the subject of

endless discussion amongst winegrowers around me. Lack of rain, too much rain, frozen rain – if only we had the dance, the stone or the stick to control it all. *Homo sapiens,* even today, will never give up trying. Too much rain can disrupt the flowering process, causing poor pollination, leading to unformed or deformed bunches. Rain is also what triggers fungal disease in the vines, and so too much rain throughout the season will make a winegrower's life hell; we try to spray when we can, often not being able to get the tractor into the vines because it will get stuck in the mud. We can only watch, helplessly, as mildew attacks the leaves and the grapes.

But drought can be even worse. Nothing can live without water. The younger the vine the shallower the root system, and the more the plant is at the mercy of drought. Old vines will resist much better but, in 2021, when Touraine experienced its worst drought in living memory, even my oldest vines struggled. We had already lost much of the harvest to spring frost, and as the hot, dry summer days dragged on and on it was beginning to look like we might lose the rest.

The frustration I felt was like that of a chained animal. Sweating in the vineyard, looking up at some high cloud passing by, I was physically willing the cloud to stop, to gather strength and release some of its precious liquid. Watching a cloud slowly drift away towards the horizon felt like being one of those shipwrecked survivors on a desert island, screaming, waving, lighting fires, as a ship glides by in the distance and slowly disappears. Over the weeks, I watched the vines start to shut down. I would take leaves in my hand and press them, trying to feel their stress, trying to picture the stomata on their undersides closing up, stopping transpiration to conserve the water within the plant. This essentially stops plant growth and, consequently, any berry ripening.

Irrigation is not allowed in most French winegrowing areas. It is seen as an anti-*terroir* practice because the grower is not respecting the *place*; he or she is using artificial means to do something that nature is meant to do, and always has. With climate change this logic is being questioned and, in some of the hardest-hit areas in Europe, vineyard irrigation is allowed. But it is very expensive to install these systems, and, when you take the problem to its logical conclusion, if there is no water coming from the sky, at some point there is no water anywhere, so how can you irrigate? Many do not see this as a sustainable answer. So, in addition to the hail rockets, there is a burgeoning industry around cloud seeding. Aeroplanes fly into clouds and seed them with droplets of a reagent designed to bring moisture together into beads of ice that will eventually fall as rain. Alas, once again, these efforts to trump nature have had very limited effect. And still we try.

That summer of 2021, when we thought there would be no harvest, the rain finally did fall. The berries were small, as we thought they would be, but the juice was good. Then last year, we lost 80 per cent of our crop because of the highest spring and summer rainfall in living memory. As these 'highest and lowest in living memory' events become more common, I don't know how farmers will cope. Maybe we can do without wine, but what will we eat? There are many people, much more expert than I am, trying to address these questions. Meanwhile, as we wait on this small farm in the centre of France, we are feeling unsure about the future. Agriculture is so important to the economic and social fabric in our region, it is not just the farmers who are worried. All our lives are uncertain.

Our farm is in the *commune* of Vernou-sur-Brenne, a somewhat sleepy place of some 2,700 souls. Vernou is like hundreds of other French villages dotted across the landscape, at the centre of which is the church, of course. Across the street, in perfect physical and philosophical juxtaposition, is the *Mairie*, or town hall, the ultimate symbol of the lay Republic, with the *tricoleur* always flying. I find something reassuring about this clear and present reminder of the French mania for complete separation of church and state.

It's odd given the size of the village, but quite in keeping with French priorities, that we seem to have two of everything else – two *boulangeries*, two *cafés*, two *charcuteries* – except, thankfully perhaps, for the one *pompes funèbres* (undertakers). The word *charcuterie* is another that doesn't really translate into English very well. It will come up as a delicatessen or pork butcher's shop, but these don't really capture it. When I think of a *charcuterie*, I can't help but think of the common rhyming French expression in my head: *Dans le cochon, tout est bon; in a pig, everything is delicious.* A charcuterie is a shop in which every single thing in a pig is transformed into something edible, all on site, and put on sale. The astounding range of pork products one finds there includes every type of ham imaginable (boiled, dried, smoked, spiced), pork chops, potted pork, trotters, sausage, pork belly – not a single piece of the pig is wasted. I will never forget the culture shock of seeing Juliette gnawing on a breaded pig's foot for the first time.

Everyone in the village has their preferred bakery – like a good-natured tribal identity, you are either a *Huvet* or a *Martin* customer. The debates among the villagers over who makes the best *croissant, tarte aux fraises* or *pain de campagne* are intricate and endless. In a world that gets noisier and more public with its opinions, the *boulangerie* debate is one of the more worthy discussions worth hearing out.

Every Thursday in Vernou, there is an outdoor market in the church square. The usual suspects can be found: a butcher, fishmonger, fruit and vegetable stalls, cheesemonger, an oyster stand. But alongside you can also find more eclectic goods: seeds of all sorts, knives, clothing, shoes, and even a fellow who will reweave your chairs, should you still have woven cane chairs. But the main attraction of the market for most people is the sheer social occasion of it all; the stories exchanged far outweigh the produce sold.

There is, in fact, something vaguely communal about living in the *commune*. You will always run into someone you know when picking up your baguette, and you are obliged to stop for a friendly chat. But it takes some time for newcomers to be taken into the fold. When I first arrived, the reticence was even more pronounced because I was a foreigner. But as a *vigneron*, a winemaker, I have been accepted, because growing vines and making wine in this part of France is a badge of honour. The activity goes back to the very founding of the first settlements in the area, and the economic and cultural importance of winegrowing over the centuries has etched itself into the common psyche of all who live here.

Now, when I go to buy my bread or oysters or potted pork, I am welcomed as someone vaguely important, at least worthy of respect. Yes, we speak about the weather, but not in the chit-chat way people usually do. They speak about it knowing the importance it has for the vines and the grapes, and the impact it will ultimately have on the harvest and our shared livelihoods. They ask if it hasn't been too dry for the vines, or if the foul weather has led to mildew on the grapes. They empathise with pruning during a week of non-stop rain, or of leaf-thinning in heavy heat. And without fail, as I walk out the door, I hear the familiar phrase, *Bon courage*, and it puts a little lightness in my step.

Every year, next to the church, the flea market springs up in the village. Inevitably, there is at least one stall selling old postcards, and I relish flipping through them, time-travelling from the late nineteenth century into the early twentieth. These shoeboxes filled with cards are like time capsules. I particularly love the ones which are simply photographs of life on the farm. You will always find cards showing people harvesting lavender in Provence, milking cows in Normandy, herding sheep in the Limousin, and naturally, a group of grape harvesters in the Loire.

These cards often have more women than men in the picture, wearing dresses of all sizes and shapes. One has women on a mountainside holding spindles, a flock of sheep behind them, with a caption that reads *I tend my sheep while weaving their wool.* Another has a woman and her young daughter in dresses in the farmyard, tending to the animals, surrounded by chickens, goats and geese. *Noah's Ark*, reads the caption. Early postcards were so often pictures of people going about their daily business, rather than pictures of tourist attractions. Perhaps the attraction was the people and the differences of the lives they led.

I wonder to myself if, one day, Juliette will start harvesting in her white cotton dress. And maybe I'll get one of those broad-brimmed straw hats that the men wear in those sepia-hued pictures.

One warm August day a few years back, a car pulled up in the driveway and out stepped a lanky, somewhat dishevelled but eminently amiable-looking man. He excused himself for bothering me and explained that he was a retired history teacher, and that his hobby was metal detecting. He pointed rather timidly towards the detector laying across the back seat of his car.

The fact that the village has been here for so long, and the land around it lived and worked on for so many centuries, makes it interesting for archaeologists and collectors of antiquity, including the hobbyists with their metal detectors. Of course, this is true for much of Europe, but my experience of it is here. My sense of the past of this place only stretches back a few hundred years, to horses and carriages, Napoleonic armies, to candles and oil lamps, wells for water and the bell ringing across the fields at mealtimes. All these things I can still almost see, and can certainly feel. But the sight of Roman legions marching past, Gallic villagers fashioning pots out of clay and artisans creating mosaic-tiled floors, are in the realm of fantasy for me, too distant and unreal.

Even so, I came a little closer to being able to picture this place in Roman times when Monsieur Gabin, the slightly eccentric teacher, spent several days over the course of a few weeks with his metal detector, combing the vineyards and the woods. He would regularly dig up unrecognisable bits of metal, mostly pieces of some agricultural machine or implement less than a hundred years old. One of his more common finds were sewing thimbles, which fit neatly into my nineteenth- and eighteenth-century sense of the place. I could easily picture women and children sitting by the fields, repairing clothing, or the men who would sometimes wear thimbles to protect their fingers from cuts while gathering straw. He also unearthed metal buttons and cufflinks, even a little chain that would have been used to fasten a cloak around the shoulders, all of these no more than a few hundred years old.

This was until Monsieur Gabin came back from the field, rubbing something in a cloth, cleaning off the soil. He laid a coin down on the table and took out a magnifying glass, peering at the piece of metal. A smile brightened his face. He looked up and said to my youngest daughter, who had been following him around eagerly most of the day, 'It's a bronze Roman coin, called a sesterce. It was minted in Rome under the Emperor Trebonianus, around 250 AD.' The wonder in her eyes captured perfectly how this object, unseen for over 1,500 years, added a deeper layer of the past to our home.

If people ask, as they often do, what our favourite vintage is, I reply that it's an impossible question. It may sound a bit odd, but to my mind it's like asking which of my children is my favourite. I may be closer in temperament to one child, or there may be times when I take more pleasure in being with one than the other, depending on what we are doing, my mood, and any number of things. But my love for each of them is the same, because my children are incomparable. It is likewise with wine. Each vintage is dear to me. There are times when one vintage might taste better than another to me, depending on the occasion, the season, my mood or what I am eating, but each one is unique.

Winemakers often have an ideal wine in mind, some perfection they are striving towards. But when I try to think about the ideal wine I'd like to make, I draw a blank. Other people have favourites, and I've often been told that they prefer a certain vintage over another. When I consider why this isn't the same for me, the only conclusion I come to is that my priority has never been to make wine. It is not to get wine awards or laurels from critics, or even to set out to make my ideal wine. There are so many wines from our vines that I have tasted and loved. They are splashed across a canvas far too wide for me to even see the end of. When I close my eyes and concentrate, trying to imagine the ideal wine, the first thing that comes to mind is the Adagietto of Mahler's 5th symphony. Could that be a wine? No, I didn't come to this vineyard to make wine. I came here to *live*. Wine is a byproduct of a desire to recover something true and basic within myself. That some people, many people, like our wine is a bonus, the cherry on top.

Not long ago I got a message from a fairly well-known sommelier in Belgium. This is a person who has studied wine in its minutest detail, whose very livelihood rests on having tasted thousands of wines and kept them all in his sense-memory, a person who spends his days and evenings recommending and serving wine to customers in an esteemed restaurant. He wrote:

'We served your 2014 and it is the best wine you have ever made. It is made for the gods!'

I don't know why at that moment this particular wine evinced this response, but I felt the way I often feel if somebody congratulates or compliments me on something one of my children has done. *How lovely to say that, but I really didn't have that much to do with it.* I was just there to guide, make suggestions. My child is his or her own person. I probably learned a lot more from them than they learned from me. I certainly owe them a great deal more than they owe me.

It takes two years to nurture the vines and grapes into wine, waiting patiently, anxiously, for juice to transform. Towards the end of August, following our tasting earlier in the summer, it is time to start bottling. Each barrel contains wine from a specific field on the specific day it was harvested, bearing the chalk-scrawled marks *South, Sept 15, East, Sept 20,* etc. The wine in each of these barrels is unique, and discovering what the final blend tastes like is another moment of excited anticipation.

Juliette and I take a day at the end of August to taste each barrel individually for the last time, before blending them together. We taste as we go, syphoning a bit of wine out of each barrel into a carafe with a specially designed glass pipette. If there are barrels that have the right profile to make an excellent champagne-style sparkling wine, we take them out of the blend and taste again, to see if taking them out has a negative impact on the final flavour. If there are any barrels we feel don't contribute to the complexity and aromatic palette, we can sell them to the local cooperative. Smelling and tasting come together at a unique time, a process of learning and revelation every year. It also takes us back through the growing season, journeying into the soil in the fields and the sap of the vines, from the broad green leaves and those ripe, sweet grapes, to the flowing, freshly pressed juice to the gentle bubbling fermentation in the barrels. We remember the heat and sun of the last few months, the rainfall and cloudy days, the wind and the calm, the morning chill of the days before harvest, even the hail and frost or prospect of drought. All of these layers of memory are rewritten into the taste and texture of the wine.

Then, when Juliette and I have decided on the final blend, we pump the contents of the barrels up into vats in the winery. Here, what we have created in the carafe in the cellar can be done on a much larger scale in these voluminous vats: a vintage ready to be put into the bottles, its glass sanctuary until the popping of the cork, somewhere off in the future, somewhere in the world.

And when bottling day finally comes, Juliette and I are always as excited as kids at Christmas. We gladly hire a subcontractor to bring in a machine which gathers the bottles on a conveyor belt, flips them upside down, rinses inside, fills them with wine, puts the corks in, and sends them back out on another conveyor belt. It's mesmerising. The best of human ingenuity. It reminds me of Charlie Chaplin's *Modern Times*, with its shiny stainless-steel nozzles, moving belts and hoses all clinking, clacking and humming. A miracle of modern engineering. Juliet and I watch the process as if staring at the animated puppets in the display windows of *les grands magasins* in Paris.

Large wineries have their own bottling machines, but they are far too expensive and take up too much space for a winery of our size. And despite all that the machine can do, people-power is still needed. There is the operator of the bottling machine, of course, who is a trained technician, and we always have five or six additional people, sometimes family and friends, sometimes seasonal workers who have helped with other tasks throughout the year. While some of us take the empty bottles, stacked high on 1000-bottle palettes, and put them one by one into the bottle-holding pen at the entrance to the bottling line, two others are stationed where the filled bottles come out on the conveyor. From here, the bottles are laid on their sides and stacked in 500-bottle metal cages. As these cages are filled, I pick them up with a forklift attached to the tractor and bring them to the entrance of the cellar, where another person takes them on a smaller forklift to position them deeper in the cellar. We won't move the bottles again for at least another three to six months, giving the wine time to settle down and gain a little more age and maturity before releasing it into the world.

Bottling takes all morning. Only when the last bottles are tucked safely in the depths of the cellar do we gather for lunch with the team. We open the first bottles of the new vintage, even though, for me, it isn't really ready for drinking. It is still wine, though. And it is good wine that everyone around the table has played a key part in making. So this is a moment for celebrating each other, not the wine itself.

Later in the afternoon or evening, when everyone is gone, Juliette and I go back down into the cellar. We take a moment there, underground, alone with the newly bottled vintage, and thank our lucky stars that all the work over the year leading to that vintage has been brought to a successful conclusion. Juliette chooses the music that we will be playing to the wine that year, and turns it on. I look at the many thousands of bottles, cool to the touch, lying on their sides, flush with the brand-new corks which will let the wine continue to breathe throughout its lifetime. There is a sense of accomplishment here in the cellar. While I relish so many of the activities outside in the vines, I still feel a deep satisfaction here, the two of us sharing this moment with the music and the wine. This is a job well done, complete, something I never really felt in my previous life.

But there isn't much time to sit back and rest on our laurels. Now the barrels are empty of wine, it is time to clean them – an early warning that summer will soon be over, as this is the first step towards preparing for the coming harvest. It always sets my nerves humming. The thought that soon, very soon, we will be pouring new juice into these clean barrels is invigorating, which is a good thing because we will need all our energies for the task – an empty barrel weighs about 50 kilos, and manipulating them requires some strength and stamina.

A few days after the bottling machine has left the farm, we start bringing the barrels out of the underground cellar, tilting each one carefully on its top edge, rolling it to the cellar door, where it can be hoisted onto a forklift and lifted up to a platform at ground level. This we repeat thirty times, positioning all the barrels in the cleaning area for their moment in the sun. Because we don't use any chemical cleaning agents, only hot water and steam, we need to be especially vigilant to be sure we get at any bacteria or fungal spores that might be clinging to small cracks. First, we scrub and rinse the outsides of each barrel with water and a brush, after which the serious work of cleaning the inside begins.

If you think about the fact that you need to clean the entire inner surface of the barrel, and can only access it through a small hole about 5cm in diameter, you can imagine what a fiddle it is. Barrels used to just get filled with hot water poured in through the bunghole and swished around. The problem is, a layer of tartrate crystals naturally builds up from the juice during fermentation on the inner walls that won't budge with just rinsing. This layer risks hiding unwelcome bacteria. So, in a compromise with technology, I use a pressure washing cane. This is an ingenious device made of stainless steel, which, unsurprisingly, is comprised principally of a cane-like steel tube. This tube is inserted through the bunghole of the barrel. I then connect a high-pressure hot water pump to the cane. When the high-pressure pump is switched on, the cane's rotating spigot sprays jets of very hot water around and around against the insides of the barrel, covering every spot. At the same time, the cane pumps the used water out, along with any tartrate residue from the sides of the barrel. This usually suffices to ensure a good clean barrel for the next vintage. However, just to be safe, every two or three years we steam-clean the barrels, which involves renting a barrel steamer.

All of this may sound like overkill, but in a winery that uses barrels to ferment and age the wine, the barrels are a crucial asset, and one of the key determinants of the characteristics of the finished wine. Even if we aren't looking to give the wine the 'oakiness' that new barrels can impart, the time spent in the barrels leaves an indelible imprint on taste, making the wine smoother and rounder. Barrel-aged wines are also more stable and usually exhibit greater aromatic complexity than wines made in stainless-steel vats, which, these days, comprises the vast majority of wines.

As summer ends, I can be found in the cellar mostly, pushing the clean barrels back into place on their stands, aligning them carefully, looking for that elusive symmetry, and putting their stoppers in. They are ready for the juice. But are the grapes ready to be pressed? Only autumn will tell.

L'AUTOMNE
AUTUMN

In the presence of nature, a wild delight runs through the man, in spite of real sorrows. Nature says — he is my creature, and maugre all his impertinent griefs, he shall be glad with me.

RAPLH WALDO EMERSON

It was one of the clearest autumn mornings I had known. The light was intense, so crisp and sharp that the headstones and tombs became translucent. I felt like I was going to faint in the glow.

The Vernou cemetery is a gem of tranquility, sitting atop a hill looking down on the village and the Brenne River. But on this day, hundreds of people were spilling from around the temporary pavilion that had been set up for the ceremony and in which François Pinon's ashes stood on a table. The people had come from far and wide, from beyond the Loire Valley, from beyond the borders of France. Virtually the entire village was there. If you came by car, you had to park some distance away and walk the rest.

Without any warning, François had had a massive heart attack. This man, this winegrower, who had been a beacon of inspiration and learning for so many of his younger colleagues (and to this somewhat older one) had vanished. It seemed impossible. It felt like the earth had opened below us, swallowing the entire Vouvray *appellation*.

The call came from his son, Julien, and I'll never forget the chill in my friend's voice. *Peter, mon père est décédé.* There were no words after that, just his sobs. I couldn't speak either. I wanted to say something to console him, but my body was rigid, my insides gripped in a burning vice. What could I possibly say that would take away his sudden grief? Through his sobs, Julien tried: 'I wanted... to tell... you,' he gasped. 'I know... you loved... him.' My sobs joined his.

A few months after the funeral a smaller, more intimate group got together down in the valley, in a field on the Pinon vineyard. Vincent had led an initiative to get a group of François' winegrower friends together, in homage and remembrance. Julien wanted to plant a tree for François, a North American maple, and we took turns digging until he placed it inside the hole and held it there while we pushed the soil back in around its roots.

Julien struggled to read a piece he had written about his father, and some of us followed with a few words. Afterwards, we walked to the winery and gathered, surrounded by vats and barrels, to have some wine. Julien came over to where I was standing with Juliette and another winegrower, and we raised a toast to his father. I asked him why he had chosen the tree that he did.

'You know, my father loved the United States,' he said. 'He loved the energy of New York and the sense of possibility. He loved the nature in America, the endless wide-open spaces, and the people he knew there. He always felt welcome, and they loved his wine! I think it was why he had a special place in his heart for you.' I didn't know what to say to that. I was so deeply moved. When Julien moved on to talk to others, I stood silently and asked myself how I came to be so lucky, standing in this ancient winery with these people.

Every autumn since, as the landscape turns from greens to browns and yellows and reds, when that sharpness of light returns, a layer of melancholy settles over me, as it does in other vineyards through these valleys, as we stop whatever task we are doing for a moment, to remember François Pinon with a mixture of sadness and pride.

When we know the grapes are a week or less away from picking, time moves quickly, at the speed of daylight, and every lit moment is filled with intense activity. The barrels may have already been cleaned, but it's essential that nothing contaminates the juice before it reaches them. Days of manic scrubbing, rinsing and buffing begins, as everything in the winery that touches grapes or the juice must be cleaned. The vats, the press, the picking baskets, the hoses, the harvesters' secateurs, even the floors and the walls of the winery. And no detergents: pure and piping hot water flowing everywhere, over everything.

There is a French proverb that says, *Making good wine requires a great deal of water.* Thankfully, this doesn't mean adding water to the juice, or even to the rainfall necessary for the grapes to grow. It refers to water for cleaning. Grape juice is sticky, filled with sugar, which means microorganisms thrive. Water, not wine, is the lifeblood. It is during harvest that I think the most about water and how much we all take it for granted. No matter how present we try to make it in our minds, and how much of an effort we make to try to use it sparingly, we still use tons of it in the winery. Literally. One cubic metre of water weighs about a ton, and the cubic metres flow freely through the ten days or so of harvest. Constantly rinsing, cleaning, rinsing again, all surfaces are done over and over as each batch of grapes comes into the press and back out. Once the winery is clean and ready for the first grapes, all eyes turn to the skies as we ask, *Will the weather hold?*

How many times have I looked to the skies since I have been here? The shape and the movement of the clouds, their constant churning, dissolving and reforming. The intensity of the sun and its changing light. But it's not just visual. You can feel the skies, the weight on your skin, the touch of humidity, the rain. And you can smell weather, too, notes of an impending storm playing in the nostrils, the earthy smell of wet soil and foliage even before the first drop falls. Before a storm, my nose picks up that sweet, almost metallic smell on the wind, which comes from ozone in the atmosphere. *Petrichor* is the word for what Juliette and I call this 'storm smell'.

The word comes from the Ancient Greek *pétra* (rock) and *ikhor*, the ethereal blood of the gods. The earthy part of this mix is caused by the formation of the volatile compound called geosmin, which is the product of bacteria living in the soil. These bacteria produce spores which are kicked into the air by raindrops hitting the earth, causing the storm smell that humans have evolved to detect. We aren't alone in being able to smell water. These geosmin signals can reveal the location of an oasis to camels in the desert, or excite fish into spawning in rising waters. On the vineyard in early autumn, the storm smell carries mixed messages. Rain is welcome after a dry August, giving the grapes that last little boost of plumpness before harvest. But if the season has already been wet, the smell triggers a foreboding, a premonition of grey rot starting before the grapes are fully ripe.

People often ask me, *How do you know when to harvest?* It's a very good question because, like all good questions, there is no simple answer. In fact, I'm not sure that there is an answer at all. If there is an answer, it might be: *I don't know*. In modern winemaking, of course, you have to know. Before a single grape is harvested, a sample batch of berries is picked and squeezed, and the juice is sent to the lab for analysis. The laboratory tells the modern winemaker the levels of sugar, acidity, pH, phenolic maturity, mix of aromatic and flavour constituents, colour compounds, and much more. Some growers use satellite imaging to give them the precise stages of maturity of every vine, based on spectrometric analysis of the foliage. And because the winemaker will already know the parameters, pre-determined by the style of wine they are trying to produce, harvest begins when the analysis tells them the grapes are ready.

But winemakers have only had access to this technology for sixty or so years. So how did they decide for the thousands of years before all this? The grower walked the vines daily, tasting grapes, using the senses of touch, taste and smell to 'feel' the fruit. The thickness of the skin, the sweetness of the juice, the palette of aromas and flavours. The phenolic ripeness can be tested by crunching some of the grape seeds between your teeth, tasting for astringency. And the sky, too, of course: the grower would be constantly reading the weather around them, perhaps using a barometer in more recent times, to help them get a sense of the risk of leaving the grapes on the vine longer. This is why, as autumn draws in, I can be found walking in the vineyard, tasting grapes, checking for any early signs of rot, trying to get a feel for when I think we should have the harvest team in the starting blocks.

The people who come to pick the grapes here are very special to me. Every year the team is different. There might be some familiar faces, they may be the same people for two years running, or there might be a completely new group. I may feel an affinity with some and less with others. I might have real discussions with some and scarcely exchange a few words with others. Some are introverts, others extroverts. Some carefree and bubbly, some serious and quiet. Some are there just to make some money, others are passionate about winegrowing. Regardless, they are all unique and important individuals to me, and it is their hands that will cut every bunch of grapes from every vine and lay it in the harvest basket at their side. They are the first point of contact with all that we have been working for, every day of an entire year. These are the people who bring our grapes home.

When you break down each task on the way to making wine into its smallest pieces, the true worth of a single bottle becomes clear. The barely audible click of harvesting shears that separates every bunch from the vine, the full weight of the bunch in the hand as the vine lets it go and gravity takes it, the curved downward trajectory of an arm as it lowers the bunch and sets it in the basket. Many people over the years have asked me how many bunches of grapes it takes to make a bottle of wine. You could ask the same question of other produce – how many apples go into a bottle of cider or fresh juice? How much wheat do you need for a loaf of bread or a bowl of pasta? How many acres of rice does a family eat in a year? How much milk is needed in all the different ways we consume dairy? But these questions also reveal just how removed we are from the farm, and how detached food is from processes that make them available on supermarket shelves.

The honest answer to all these questions is: it depends. In winemaking, different grape varieties express their fruit bunches differently. Some have berries that are larger than others, some have bunches that are more compact and tighter. But even within the same variety it depends on the variables of the soil, temperature variations, rainfall, cloud cover, the human touch.

At our vineyard, because we are driven by the principle of letting nature guide us, how many bunches of grapes we need to make one bottle of wine changes all the time. There are years when the berries are larger and juicier, and years when the bunches have fewer grapes than others. So when a basket full of grapes arrives from a row of vines and is slid onto the wooden trailer, I don't think about the weight of the baskets, how many bunches or berries it contains, or how many bottles of wine it will make. I think of the rain, the sunlight, the soil, the miraculous balance between it all and the human activities and energy that flows into growing those grapes, the hours of physicality and attention, anticipation and hope.

It is only since I have been a winegrower myself that I have fully appreciated the value of produce raised by small farms or made by artisans. The nurtured, the hand-made, the traditionally crafted. Seen through the cynical gaze of commerce, these things can come across as marketing and branding to justify higher prices. But there is much more to it, and if these things are truly authentic, then the value they have is fast disappearing along with our ability to really understand the costs. Sometimes, a price tag will never be enough.

Before I came to the land I assumed, as do many people, that all wine grapes are picked by hand. After all, if you look at a bunch of grapes it's hard to imagine it being separated from the vine in any other way. How could a machine pick these bunches of soft fruit from different locations on a grape vine? But human ingenuity is boundless, especially when business and economics are at stake. An advert on TV or in a magazine that is selling wine with an image of people out harvesting in the vineyards, picking the grapes by hand as the sun rises over verdant valleys, is very misleading. Today, roughly 90 per cent of grapes destined for making wine in the world are picked by machines. Machines are faster, cheaper and (somewhat) more reliable than people. Implacable logic.

But until the day when terminator-type humanoid robots can crouch before a vine and snip off bunches, judging their ripeness and quality with sensors which replace human eyes, noses and taste buds, machine harvesting will remain very primitive and violent. Machine harvesters are massive tractors – 4 metres high, weighing in the vicinity of 5 tons, with engines of 150 or more horsepower. They look like something designed to mine uranium ore on Mars. The machine straddles a row of vines, with the driver encased in a glass cabin above it all, and the beast growls down over the row of vines as a series of rubber batons beat the vines, shaking the berries chaotically off the stems so they fall onto a conveyor belt which transports them to the holding bin. When the bin is full, the machine dumps its harvest into a large, stainless-steel receptacle waiting at the end of the rows, which, when full up, is taken to the winery so the grapes can be pressed. Job done.

But wait, let's think about this machine process one bit at a time. Extremely heavy machines with massive tyres compact the soil as they roll through the vineyard, causing asphyxiation of the ground that the vines grow in. Powerful engines burn large amounts of diesel, adding to the pollution caused by industrial agriculture. Batons hit the vine with such force that not only to the berries separate from the stems, but so do leaves, twigs, dirt and insects living on the vine, which means much of the crop is broken in the process and the juice is exposed to the air – this starts oxidation and bacterial activity as the grapes wait, often in the hot sun, to get to the cellar. Of course, a machine cannot (yet) distinguish between bunches that are unripe or have started to rot, as a human picker can. And because you get a lot more than ripe grapes in the mix, it means that the juice will need 'correcting' later on, using modern oenology, so it meets the parameters that the winemaker is aiming for and makes it suitable for drinkable wine.

Because I see a vine as more than just an entity of economic production, the idea of beating a vine with rubber batons makes me cringe. Using a machine in our vineyard would be impossible because old vines can be very badly damaged by the intervention. That's why industrial vineyards are uniformly planted on plots as flat as possible, and the vines are ripped up and replanted while they're still young. Fortunately, in some parts of France, vines have been planted for centuries on terrain too steep for machines.

One of my favourite times during the harvest period is the early morning. We can't start picking until sunrise, when there is enough light to be able to work, and I am always up an hour before. The mornings are cool, sometimes downright cold. The layers of clothing I start the harvest with gradually get peeled off as the day progresses. I head out to the winery with my coffee and put some music on, always something peaceful, perhaps a string quartet, and I sit in front of the press and picture the day ahead. I anticipate the empty vats being filled with juice. I visualise the parcels of land and the vines we'll be picking, wondering about the pace we should move at, depending on the health of the grapes and the day's forecast, asking myself questions: *Will we pick everything in the parcel? Or will we select bunches and then wait a few days to go back for a second pass?*

Some harvest mornings, sitting in the centuries-old winery, I'll get up and go over to the wall behind the grape press. Here, over the years, people who have worked in the winery have carved their names into the *tuffeau* stone. I run my hands over the names: Emile, Marcel, Jean. The earliest etchings go back at least a hundred years. I haven't carved my name into that stone. I don't know why. Perhaps I don't quite deserve it yet. Perhaps I feel in some way that if I do I will become part of the past of this place rather than its future. I do know that on some of those solitary mornings in the winery, as I wait for the harvest team, I feel that the time to carve probably isn't all that far away. At some point, sooner or later, my name will be there among the others.

All those years ago, when we moved into Le Clos de la Meslerie, the winemaking equipment was defunct. The cellar was without electricity. The barrels were rotting, some half full of wine which had turned to vinegar. There were broken bottles, remains of old corks and hoses, manual corking machines and pumps were scattered about, untouched since the last day wine was made here decades earlier. The earthen floor and stone walls were heavy with the patina of the generations of people involved in making the wine throughout the farm's history.

Above the cellar, the winemaking building was piled to the ceiling with furniture and other discarded remnants from the house and its history. As we gradually cleared the cellar and the winery, we discovered several full bottles dating back to the 1960s. Best of all these discoveries was the old press. We didn't know it was there when we moved in. It was hidden entirely from sight, and we had assumed that buying a new, pneumatic press would be necessary to resuscitate winemaking at the farm. These electric presses are ingenious: a large stainless-steel canister is at the heart of the machine, inside which is a rubber bladder that inflates and deflates, powered by an air compressor. You put the grapes into the canister and inflate the bladder to press the grapes.

When we discovered the old manual press, which was manufactured in the 1920s, it was as if providence was telling us something. Was it possible that we could use this press to make wine? How uncanny, given the underlying philosophy of how we wanted to approach winegrowing, that this press was still there. Looking at the old machine, with its numerous wooden parts and its imposing, massive steel and cast-iron central mechanism, I was instantly enchanted. I also had no idea how it worked.

I called Vincent and Damien to take a look, and while they could picture it in action from long lost images in their minds, none of us could put all the pieces together. We asked Damien's father, Jackie, to see if he could help revive the press. He had worked on something similar as a boy in the early 1950s, and we spent an afternoon together as he showed me how to assemble the press and install the wooden slats on the basket. We filled the press with old sheets and blankets, to simulate a load of grapes, and put the oak slabs of the press cover on top. We then loaded the oak beams in layers – Jackie told me that these oak beams used to be called *cochons* or pigs, because they were so heavy. I laughed at the time, but over the years I have come to know these *cochons* very well, and often curse them under my breath on days when, while lifting them on and off repeatedly, I feel the burn in my arms and back.

We checked the hydraulic system and greased the pistons and massive central screw around which the entire enterprise would unfold. Finally, we wound the press body down the central screw until it sat on top of the *cochons*, tightened it as far as it would go within the limits of our biceps and triceps, and closed off the hydraulic system so that we would be able to build pressure as we worked the pump lever.

All set up like this, the press took on a reality that I could hardly have imagined when we uncovered it in its dilapidated state. This might actually work. Slowly, I began pumping the lever. Resistance began to build, little by little. And suddenly, to my delight, the needle on the pressure gauge came quivering to life after at least thirty years of dormancy. The press was resurrected. That moment, seeing the look on Damien's father's face, will never leave me. We gave it a coat of bright-red paint, and a year later performed the same series of motions as we poured in the first baskets of *real* grapes. As the bunches piled high, up to the top of their wooden-slatted vessel, and we put the last *cochons* in place, a small gathering of people formed with intense anticipation, ready to take in the spectacle. Friends, curious local winegrowers, even a journalist, looked on as I worked the lever and the first, pure, delicious juice began seeping out between the wooden slats and flowed down to the waiting vats. Finally, Le Clos de la Meslerie had risen from its ashes.

Today, we still press all our grapes with this hundred-year-old, vertical wooden basket press. If not the last, we are certainly one of the very few vineyards in France to do this. Our press has become emblematic of the vineyard, expressing our approach to winegrowing and the question at the root of our livelihood: *How did they make wine before 1950?*

After our first harvest an old-timer came up to me and said, 'That press will make a wine like no other in the region.' It's difficult, if not impossible, to pinpoint what it is that gives any wine its character, because there are just too many variables. But nobody else in the region uses our press, and it has become the common thread that runs through all our vintages. Exactly how much of it comes from the press is impossible to know, but its rediscovery and revival has become the essence of what we do. Some people have suggested that we try and find out by experimenting, pressing part of the harvest in a modern press to compare. It's not unusual to experiment, some growers do it all the time in small ways, experimenting with leaf-thinning one parcel and not another, or putting down different amounts of manure each year. And although I am a great believer in the scientific method, it does not fit with what we're doing here.

We try to make wine by feel and intuition. Inevitably, this has led us to make some rather eccentric wines, but I stand by them all without reserve. They are all part of a journey. Nor does our approach mean that I think our wines are superior in some way, but they certainly are unique. And for the first time in my life, I can truly say that I am making no compromises in what I am doing. I am under no external pressure to do things differently, to be more efficient, more productive, more aggressive, more hi-tech. Seeing people appreciate our wine, I'm reassured that taking a different approach, being outside of the mainstream, has been a risk well worth taking.

Some mornings, when autumn fog folds around the vines, you hardly see the person walking a few metres in front of you. As the harvest team heads down into the vineyard, each of us carrying six stacked empty picking baskets, we are like a troop of ghostly grape pickers. The vast Loire River is only a few kilometres away. Even closer is the Brenne, which flows into the Cisse then into the Loire. All these waterways provide the moisture for these mornings of rising fog. It is one of the reasons that for centuries this area was best known for its very sweet wines, along with Sauternes in the Bordeaux region and Tokaj in Hungary. Each of these regions owes its reputation for these exquisite wines to their privileged relationship with nearby rivers. As the warmth of the rivers meets the cooler morning air, fog rises from these bodies of water and settles on the grape bunches, triggering the development of *botrytis cinerea*, or 'noble rot'.

The traditional sweet wines, or *liquoreux* as they are known in the region, are concentrated explosions of complex flavour and aroma: honey and beeswax, litchi and mango, dried fig and apricot, cinnamon and saffron, acacia flowers and cherry blossom. These wines, to my mind, are some of the most intriguing and succulent of all wines. Whereas mass-producers of sweet wines simply add sugar, these powerful flavours come from rotting grapes. It is a beguiling paradox, and takes some courage on the winegrower's part to hold back and let nature take its course, because noble rot can very quickly turn to sour rot, which ruins the harvest if the conditions aren't right. And the right conditions for harvest means that the fog *must* lift by late morning and leave behind a clear sunny day. This dries the rot that has begun in the fog and prevents it from turning to sour rot. The warmth of the sun also encourages some of the water in the juice to start evaporating, leaving the grape through tiny holes that the microscopic fungus has pierced in the skin. This is what causes the celebrated shrivelling of the grapes, within which every drop of juice is extremely concentrated in flavour. As you can imagine, the harvest yields for these wines are minuscule, but when successful the result is exquisite. Here at Le Clos de la Meslerie, we will get some noble rot every few vintages, but I have yet to make a completely *botrytised* wine.

As we descend into the fog-covered vines at the sun's first light, the bunches draped in tiny droplets of water, I think of the time I tasted a 1921 Vouvray *liquoreux*. I remember the emotion that I felt, and the privilege I sensed at now being a part, however small, of this tradition. It rekindles my hope that one day soon we will have the opportunity and courage to make our very own *liquoreux*.

Although there is much to juggle back at the winery as the harvest comes in, I always start the day picking with the team until we gather enough grapes to get the first press of the day underway. And because I will then spend most of the day pressing, it's important for me to at least begin the day in close proximity to the grapes. It would be wrong if I didn't. Over the course of the season, so much time and passion and energy has flowed into growing them that to finally see the beautiful ripe bunches hanging on the vines gives me a huge sense of gratification and accomplishment. As I fill the baskets with 15 or 20 kilos and carry them up to the trailer, stacking them with the other full baskets that the tractor will tow to the winery, I feel a deep sense within me repeating, *This is what I am meant to be doing.*

Getting close to the grapes, observing, touching and tasting them, also gives me a much better sense of the state of the harvest. I get a feel for the 'sanitary' condition and ripeness of the vintage. There are years when these can be so variable that we need to do several passes, picking only the ripest bunches, and taking out the bunches showing the first signs of unwanted rot to avoid it growing and spreading. Being with the team also gives me a sense of the character of the people, how they work together and the energy they have, and whether I need to be more or less present throughout the day, encouraging and guiding.

As I go down each row, hearing the click of harvesting shears cutting bunches from the vines, squatting, kneeling, standing, lifting the basket to the next vine, again and again, and carrying the baskets, heavy with fruit, up and down to the end of the rows, I feel whole. The symbiosis of my brain, my mind, with my body, is complete. This is what these hands, arms, legs and feet are for. Some go to the gym, go jogging, do Pilates or yoga, go biking, because we know, deep in our DNA, that our bodies are made to be used, that without putting them to action, something is wrong. Our bodies were originally built to work to help us survive. To hunt and gather, to work on and with the land, to pick fruit.

When there are enough full baskets to make up a press load (about 800 kilos), I get on the tractor and drive the laden trailer up to the winery. Juliette and I always work together on the press during harvest, usually with one or two helpers. It's a rare moment in our life together. The year is always so busy, with me in the vines and cellar and her managing the vineyard visitors and much of the marketing for the wine. These few weeks of harvest give us a chance to spend full days together, working, enjoying each other's company and sharing the pleasure of the end result of the year's toil. This juice that we are pressing will sustain us, and allow us to continue this adventure together.

Working together, loading and operating the press, exhausting our minds and bodies, Juliette and I feel a form of transcendence. It is one of the reasons people drink wine. We leave the chores and routines of life behind and feel part of something greater than ourselves. We know that people will be drinking the wine made from this juice two years from now. After it has been born and aged in the cellar, the wine will travel the world to be enjoyed in restaurants, at dinner parties, on beaches and piers, before sunsets and crackling fires, and in some small way, as we press, we are getting a little bit closer to all those people we will never meet.

As we pump the lever, building pressure on the full press, watching the grapes being gently squeezed, the first juice begins to trickle out from between the wooden slats. Then it builds momentum, gradually reaching a flow, and we smile again and give each other a hug. My youngest daughter always wants to be first to taste the juice, so she grabs a glass and fills it with the cloudy liquid. We all taste. And with this new nectar on the tongue, filling the mouth, we register the first expressions of the vintage's character, a sketch of what the final wine will taste like. I picture to myself the bare, stark, sleeping winter vines and am filled with awe at what they have produced.

When I think back over recent harvests, one particular group of pickers comes fondly to mind. I remember Nataliya, who had studied ballet throughout her youth, rising slowly and elegantly from her picking crouch position to strike a stunning arabesque with a bunch of grapes held high. There was also sixty-five-year-old Ihor, waving me over to show me a handful of sloe berries that he had taken from a blackthorn in the nearby hedgerow, asking eagerly, childlike, if he could pick more to make jam. Then there was the look in Kseniya's eyes, a blend of joy and fear, as she told us of her plans to try to get back home to marry her fiancé who was a soldier on the front line in Ukraine.

When war came back to Europe it was a terrible reminder that history is too often cyclical, and in the wake of those cycles there is always suffering. As Ukrainian refugees flowed into France, a local website was set up to help them find places to stay in the region, mostly in the city of Tours, about 15 kilometres from the vineyard. As harvest drew nearer and we were starting to plan for the team we would need, we decided to see if there were any of these refugees who might be interested in joining us. The response was almost immediate, and one bright, warm morning we walked through the East and South plots with Nataliya, Kseniya, Oksana, Ihor, Maiia and Andrei. We showed them the vines, the grapes, the secateurs, and explained how it would all work. This small group consisted of a banker, a urologist, a civil servant, a graphic designer and an artist. They shared a common language, heritage and flight from their war-torn country, and all brought to us their enthusiasm, gratitude and willingness to share some common humanity.

As we spent more time with our new coworkers we learned

that, despite being grateful to have been taken in by France and given food and shelter, they were feeling listless and depressed. Their lives had become taking French lessons, navigating the bureaucracy for refugees, and getting increasingly bad news from their families and friends who remained in Kyiv. They were thrilled to have the opportunity on the vineyard, not just to make some money but to get out and do something. More than anything, they wanted to be useful.

Despite being city-dwellers in Ukraine, farming and the land were still close to all of them, and living on the land was in most cases only one generation away. If there wasn't still a farm somewhere in the family, they had a country cottage where they grew fruit and vegetables, and plant-based home remedies still played a big role in their lives. I often spotted Ihor, the urologist, looking curiously at various plants growing in the vineyard or on the hedgerows, smelling and even tasting them. It soon became clear to me that we all shared a common language: a love of the land and farming.

We spent ten days together in total, working, resting, eating, drinking, and talking as best we could. They did not want to talk about the war. They wanted a moment of respite from it and to have something that resembled normal life. They wanted to tell us about their country, their families, their lives, and to learn about ours. They told us of their cuisine, their music, and of course, their wine, which had been part of their culture for centuries.

When the harvest was done we threw a celebratory lunch. Our Ukrainian friends came bearing traditional dishes and the wine flowed as freely as the discussions; a cacophony of French, English and Ukrainian swirled around our long outdoor table. After a group toast to eternal friendship with a shot of *horlika*, a rather strong Ukrainian digestif which Ihor had somehow procured, Andrei brought out his guitar. Nataliya and Kseniya had wonderful voices, and we spent the rest of the afternoon listening to Ukrainian folk songs, mostly songs about peasant life on the land, sometimes joyful, but mostly very wistful and filled with unbearable longing. There were tears in many eyes as the words, unintelligible to me, blended with the music. Our Ukrainian team, for a moment, was transported home.

During our first couple of harvests the exhaustion I felt at the end of the day was so extreme that I wondered if we were going to be able to continue pressing manually or whether we should succumb to technology. It's easy enough picking up a 20 kg basket of grapes from the trailer, carrying it into the winery and lifting it to chest height to pour the grapes into the press. But when it came to doing this thirty-five times for each pressing, with three or four presses a day over the course of the harvest, this piece of the process alone meant lifting and carrying roughly 15 tons of grapes. Putting the oak boards and beams into the press added several more tons, and this all before reaching the most physically demanding part of pressing manually: the step known as *rebêchage*, for which there is no direct translation into English except 'breaking up of the press cake', which makes it sound more like a birthday party than the 'torture' which better describes the process.

As the grapes are pressed, a point is reached where the juice stops flowing and the press cannot exert any more pressure. This does not mean that there is no more juice in the grapes. It simply means that they have been pressed into a tight 'cake' from which no more juice will flow at maximum pressure of the press (for a manual press this is much less than for automated presses). I haven't investigated the physics of this phenomenon, but it means that the cake needs to be broken up and the grapes pressed again. In modern presses, the expandable bladder deflates and the entire press rotates until the cake breaks up. In our press, all the oak boards and beams need to be taken off the cake one by one, and a worker, or two, take pitchforks to the tight mass in order to break it up into chunks. This may not sound like much, but a grape cake is surprisingly solid and resistant. So although breaking it up sounds quite fun, the cardio and upper-body exertion is the equivalent of going several rounds in a boxing ring. Yet sixteen years after our first harvest, we still haven't been tempted by a modern press. We are still breaking the cake.

We usually have at least one other person helping in the winery, so that Juliette and I can give each other breaks, rotating the various tasks. She constantly amazes me with her strength. Every autumn, she becomes Demeter, the ancient Greek goddess of the harvest and agriculture. She is everywhere at once. She is the glue that holds the entire operation together. Strength and perseverance in the winery and at the press, encouragement in the fields, and provider of sustenance. When we go out into the fields to grab the heavy baskets filled with grapes, carry them out of the rows and load them onto the trailer, she is there. The harvest team gets energy from her presence; they look to her for direction, they go to her if they are injured or not feeling well. She is the one that brings out the big wicker basket filled with goodies to eat, along with the tea and coffee and water at mid-morning break. She is the one that offers the harvesters a bucket of hot water and towel so they can rinse their sticky hands every now and then. She is the one that somehow conjures up a meal for ten people between all the other tasks she's involved in.

I sometimes watch her from a distance, when she isn't looking, and wonder how she does it. I am so single-minded, focused so closely on the process, the logistics of the event that, without her, I would miss much of the joy of the moment, the satisfaction and rewarding human contact. We say that we have made so much progress on women' rights, and I do believe that we have. But I wonder if the women on family farms in past centuries felt they had an importance, respect, strength and sense of being essential and irreplaceable that many today don't.

If I were forced to exert myself to the same extent doing something which I didn't enjoy, I wouldn't be able to do it. My mind would rebel, my spirit would desert. But doing it as part of this vineyard's life, knowing that it is a crucial step in the process of creating the wine in as natural a way as possible, makes it not only bearable, but almost exhilarating. The exhaustion I feel at the end of a day of harvesting, as I stand under a hot shower, feeling my muscles slowly relax, the exquisite pleasure of the evening meal and wine at the end of the day, the deep, restorative slumber that follows: all of this is the reward for making the extra effort through the harvest.

It's been sixteen harvests on the old press, mostly without fail. Having been roused from her thirty-year sleep in 2007, it's as if she had never stopped, and now she's over a hundred years old. A few years ago, however, she was wounded. Juliette was pumping the lever and noticed that she was having difficulty building pressure. Mickael the Bear was beside her, hunched over the edge of the sturdy oak-slatted basket. He called me over and pointed to a small trail of hydraulic oil underneath the lever assembly. He asked Juliette to resume pumping and, sure enough, with each descent of the lever the oil surged. Drops were falling onto the wooden planks below, under which was a batch of our half-pressed precious grapes. The press had sprung a leak. The hydraulic oil in the chamber is tasteless and food-grade, therefore harmless for human consumption, but we put a container under the leak all the same.

Juliette, Mickael and I stood silent, looking at each other. A panic was rising, tears were welling in Juliette's eyes. We had a half-pressed load in the basket, at least two more days of harvest before us, and suddenly no press. Trying to calm myself and think clearly, I started to rack my brain. What were the options? I loved this press. It was a piece of history and part of our folklore now, but history and folklore gave cold comfort in this modern emergency. Our local agricultural mechanic, Michel, had told me years before that he wouldn't work on the press. He had been a lifesaver many times over the years when I had problems with the tractor, the sprayer, the mulcher, the ploughs. I had asked him once if he thought he could service the press as I had assumed that, like all machines today, a service to replace worn out parts was probably a good idea. He smiled and said that he wouldn't dare go near it. As far as presses went, he could repair failed electronics, faulty electric pumps, broken pressing bladders and rotating motors. But he feared, if he started fiddling with our press, he might break things that could no longer be repaired or replaced. It was impossible to find spare parts. 'Use it until you can't use it anymore,' was his advice. Was this *that* moment? When we couldn't use it anymore?

In the end, what saved us was the solidarity of the people in the farming community. The first call I made, as usual, was to Vincent. He was in the throes of his own harvest but he dropped everything and arrived. No way to repair the press now, he said. We could unload the half-pressed bunch and drive it down to his winery a few kilometres away, where he could press it for us. But his press was too large to press small batches like ours. He would probably have to mix our grapes in with his, and just give us a percentage of the juice, which we would then have to transport back up to our winery. It could probably be done but the logistics were very complicated, and a part of our juice would not be from our land.

'I have another idea that might work,' he said, taking his phone out of his pocket and tapping the screen. He put the phone to his ear. 'Christophe, I have a problem.'

Christophe was another Vouvray grower. I had met him in passing but never really spoken to him. Vincent remembered that a few years back Christophe had purchased an electric, pneumatic mini-press to press small batches of late-harvest grapes in years when it was possible to make the famous *liquoreux* wines. Christophe said that he would be happy for us to bring our harvest to his place but that he knew it would be easier and more comfortable if we had the press at our place. He had a truck big enough to carry it, but would need help getting it onto the trailer as it weighed almost half a ton. I grabbed Sergey, one of our Ukrainian harvesters with a broad set of shoulders, and he joined Juliette, Mickael and me as we tore off to Christophe's winery.

Christophe's machine saved the day, and we finished the year's harvest in peace. We are forever grateful to him. But we still had the problem of what to do about the old press. Was this part of our story finished? Would she just become a museum piece, sitting there to show visitors how things used to be done?

On the last day of that fateful harvest, as we toasted with the harvest team, I noticed Sergey wandering into the winery and looking at the press with what seemed to be more than simple curiosity. When he climbed into the press and started to inspect the lever mechanism, from where the leak had sprung, I went over to him. He had caught my interest over the course of harvest. Middle-aged, tall and strong, he had left Kyiv with his young family before they stopped letting the men out and sending them all to the army. Even though he was taking the French classes that were obligatory for Ukrainian refugees, he still couldn't really speak a word of it. He often seemed a little dejected, staying slightly apart from the group, smoking sullenly. I don't think he particularly wanted to learn French. He just wanted to go home.

Sergey climbed out of the press, took out his phone, and spoke into it in Ukrainian. He handed it to me. Google translate showed me words that sent a frisson of delight through me. 'I think I can fix this.' Through the phone, Sergey explained that in Ukraine he had been an industrial mechanic. He had spent his life repairing machines of every shape and size. My delight wobbled though when I thought about what Michel had said about his reluctance to work on the press – if anything went wrong, we wouldn't be able to get any parts.

I took the phone from Sergey and said, 'But this machine is a hundred years old. Isn't that a problem?' For the first time since he had been with us, a broad smile broke on Sergey's face. He spoke into the phone and handed it back to me. 'All the machines in Ukraine are a hundred years old!' The two of us broke out into laughter, which brought others in to see what was so amusing. It was the start of a friendship that I value to this day.

The following week, he was back with his massive toolbox, taking apart the press. I stayed with him, curious, helping where I could, as he wrenched open bolts that hadn't moved for a century. It took some time, but Sergey is nothing if not patient and he found the faulty gasket, prying it free with a long screwdriver. He handed it to me, victorious. It was of a soft material, concave in shape, unlike any joint or seal I had ever seen. Sergey smiled at me and handed me his phone. 'It's made of leather. Most seals were before rubber and plastic became commonly available.'

I would later learn, as I combed the internet trying to find a replacement, that leather is still used to make seals and gaskets for certain hydraulic machines. Its strength and flexibility will often do the job better than any artificial materials. The problem was finding this exact model, with its unique size and shape. Sergey and I had another laugh when he pulled up a picture on his phone showing what looked like the identical gasket. Excited, I asked him where the store was. Could we go there? Could we order it online? 'Well,' he wrote. 'It's in Smolensk, Russia.'

Eventually, after several weeks of searching, we found a supplier in Romania. When I ordered the part, I asked Sergey if I should order two just in case. 'Only if you think your great-grandson might need one,' he replied with a gleam in his eye.

As the years pass and the number of harvests mounts, I often find myself thinking back to our first harvest in 2008. With sixteen harvests behind us, our remarkable naiveté that first year seems, in retrospect, both exhilarating and frightening. Yes, we had, as the song goes, 'a little help from my friends'. Maybe even a lot of help. But, in the end, we still had to leap into the unknown. It was up to Juliette and me to figure out the nitty gritty of it all.

We wanted to start off gradually, especially since we didn't really know what to expect from the press. Vincent had agreed to purchase the grapes from North and West plots, so we didn't have too much, leaving us with about half of the harvest. This took some of the pressure off. We could take our time, if the weather allowed. We had found (again, through Vincent's network) a group of students for the harvest team, all of them doing a master's degree in winegrowing in Angers, a city a few hours from here. We would provide room and board, and they would provide so much more than I could ever have imagined. It was an international group: three from South Africa, one from Mexico, one from Croatia and another from Brazil. They all had experience working on vineyards, either during or in between their various degrees. Which meant, combined, they had a good deal more experience and knowledge than I did.

Throughout the five days that we harvested, these 'kids' showed so much enthusiasm, diligence and knowledge. Their youthful joy was infectious, and it set the tone for the whole experience. Our youngest, Célestine, wouldn't be with us to taste the juice for years yet, but Daphné and Alexandre, six and five at the time, looked up to this picking team like gods, following them around in the vineyard, playing hide and seek, frolicking to the spirit of the day.

We got lucky with the weather, and the season in general had been a good one, so the grapes were beautifully ripe and had no sign of disease or pest damage. As batch after batch of grapes came into the winery, Juliette and I got acquainted with the press. We filled the basket to the brim with the golden grapes the harvesters brought us. Seeing that basket full of close to a ton of delicious fruit, still cool from the dew-covered morning, felt like reaching a summit. The intense climb of the year, ploughing and pruning, destemming and leaf-thinning, trellising and mowing: it was all for this. I took a photo of the filled basket and newly painted red press. We covered the grapes with the oak boards and beams and screwed the press body down its axis as tightly as possible. We were ready.

Seeing the fruit in the press basket was only one of the many firsts we would experience that day. The anticipation of tasting the first juice was almost unbearable. With the press ready to go, we decided to call the harvest team in from the vines for the occasion. Each of them took turns pumping the lever to build pressure. Juliette and I went last, and the first trickle turned into a flow coming from between the slats of the grape-filled basket. As we took glasses and tasted the juice, I thought of all those times over the past century that the juice from the grapes grown here flowed through those slats, and tried to picture some of those people standing around the press like we were, tasting the juice that would become their wine.

Juliette and I had no idea if what we were tasting was going to be any good. It would take years of experience to be able to taste the newly pressed juice and be able to have an inkling of what the wine would be like. This time, we just knew that the juice was delicious. Daphné and Alexandre agreed, as they crouched over the basin the juice was flowing into, dipping and licking their fingers while saving the occasional ladybird or small spider that had inadvertently been swept into the press from drowning in this sweet, sticky nectar.

That evening, after finishing the day with several hours of cleaning in the winery, we put on a celebration dinner for the team and ourselves. We didn't know how exhausted we would be when we had planned the celebration, but we were running on adrenaline by the time we brought out a huge pot of stew, piles of bread and bottles of wine. We had invited friends as well, and after dinner we put music on in the courtyard and some of us danced. At one point, I saw Juliette and Vincent standing apart talking. I wandered over to thank Vincent again for all he had done to help us get to this point and, as I approached, I saw that his eyes were glistening with tears. I wondered what could possibly be wrong. But then I remembered that this was Vincent, who had helped, encouraged and supported us in the journey to get to this day. These were tears of happiness for us.

The last day of every harvest brings a similar mix of emotions. While the sense of achievement is great, and there is even an element of relief to have got through without any major disasters, it is still the end of something unique and life-enhancing. Having spent such an intense time with the team of pickers, those helping in the winery and others who have stopped by to help or provide moral support, you get to feel a part of something greater than yourself. While we may be the owners of this vineyard, I feel nothing of it. We are simply part of a group effort to get something done, a task which seems to me both uplifting and worthwhile.

People often imagine that there is a big party every year on the last evening of harvest to celebrate, but this is rare. Everyone is just too exhausted. Usually, winemakers will have the end-of-harvest dinner or party a few days or even weeks later, once everyone has rested up. For Juliette and me, the last evening is almost sacred. When the final press has been filled, around mid-afternoon, we bring the team together, open a few bottles, raise a toast, thanking everyone profusely for their help. We chat for a while until, gradually, wearily, everyone disperses.

Then the two of us put music on and take our time pressing the last batch of grapes. We savour it. We make it last. The next step is several hours cleaning the press and winery, a task that can be danced through with a bit more music and wine.

The word *cynefin* is layered with meaning. A Welsh noun with no direct equivalent in English, its origins lie in a farming term used to describe the habitual tracks and trails worn by animals in hillsides. The word has since morphed and deepened to express a very personal sense of place, belonging and familiarity. The artist Kyffin Williams described *cynefin* as a relationship, 'the place of your birth and of your upbringing, the environment in which you live and to which you are naturally acclimatised.'

Before I came to the vineyard and started living within it and its rhythms, I had never felt anything like *cynefin*. Being 'naturally acclimatised' to a place is, I think, only really felt by living within nature. Over time, I began to understand *cynefin*. For me, its meaning is found in the natural topography of my surroundings and how the meaning of a place is intimately linked to its shape. It was years, maybe even a decade, before I became physically and mentally attuned to the shape of the land around me, and its contours became a part of me. Today, I don't just live in a house. I live in an ecosystem which is shaped by climate, geology, history, the work we do here and *topography*. Existing on the top of a hill is different from existing on a plain, or in a valley. Existing surrounded by land carved from the millennial flow of rivers is different from being on land flattened as far as the eye can see by the scrape of glaciers. When I lived in cities, I was completely ignorant of this interplay between the mind and the topography of the place. Visiting the mountains or the sea is one thing, but it takes time to fully sense the lay of the land.

When I am working on the steeper slopes of the North plot,

with the woods below, I have a different sensation from when I am at the bottom of the plot, with the trees rising above, or when I am in the West plot where the flat of the hilltop stretches out before me. The openness of the one contrasts with the close compactness of the other to evince a different sense of perception, and perhaps even wellbeing. Several hundred thousand years of evolution of the brain make the exposure of being on a long, flat stretch feel very different from being tucked safely between two hills and the woods.

Topography also plays hide and seek with its residents. There are many places I can see or check on from at least one point in the house. There are other places that I can't see unless I actually go there. Those places usually harbour activities that wouldn't be happening in the places more exposed to my vision, like the resting deer, nesting birds or rutting boar. There are days when Juliette and I walk the entirety of the perimeter of the vineyard, just to feel these gently shifting sensations and moods. Every slope, every flat, every angle is familiar now. I notice the action of the rain and the wind, the dying and regrowth, the falling of branches and growth of trees. I know that just beyond that rise there is a hollow where the grass will be flattened from sleeping deer the night before, or that after the curve in the path I'll come across an abandoned foxhole in the side of the hill, which, in a few years, will be filled in by stones and earth moving down the slope. I know where, along the endless hedgerows surrounding the vineyard, the slope of the land creates pockets where the most rainwater gathers, offering up the sweetest, ripest wild blackberries and plums. And this intimacy keeps feeding and shaping our lives here.

Ever since I became a father, I have wondered what it means to be a good one. Like most parents, I am sometimes racked with guilt about not being present enough, not giving enough of my time and energy to my children. Running this vineyard is so much work, so time-consuming and mentally absorbing, that I have struggled with giving more of myself to my children. There is no doubt that I could have done more, but as I watched them growing up in this place, *cynefin* feeding them as it has fed Juliette and me, I know that I could have done a lot worse.

Because we live where we work and work where we live, I was never away on business trips or returning from the office after the children were already in bed. I've never been so stressed by the pressure of my job that my family suffered for it, all things which would have been inevitable if I had remained in my previous life. Yes, I work a lot and often on holidays and weekends, but I am always nearby. They can come to see me when I am pruning or leaf-thinning. They can hop onto the tractor with me when I am ploughing, and I can give them a quick hug when I come in for a break.

It is true that had I remained in my old world I would have been able to give them more materially. Maybe I would have been able to go skiing with them in the Rockies, or taken them on yachts or to resorts in the Maldives, or walked the Great Wall of China together. If things had gone according to the plan, I could have kept them in the latest designer fashion, sent them to fancy summer camps and to boarding school at Choate or Harrow. But just writing these words makes me see how absurd they are.

Any doubts I have melt away completely in the autumn. Starting from when they were barely toddlers, they have always joined in the harvest excitedly, picking with the team, trying to pull the lever on the press, grabbing a glass to taste the fresh juice. My eldest two are already off at university now, but I believe that this place and this life have left an indelible mark. I'm convinced they feel deeply the meaning of what we have tried to do here. If, as they venture out into the world, what we have done here has given them even a hint of the joy, peace and satisfaction that one can find in work, if it has brought them a little closer to the natural world and the profundity of its diversity and cycles, if it has made them feel a little bit more an integral part of this planet rather than an accessory, then that is something.

A few years ago, we decided to add other family lives to this place: we invited Juliette's parents to come and live here. They hadn't asked for it but Juliette and I both saw that it was time for a change. They knew, too, that they'd reached an age when they needed help with some of the basics. Juliette's father was starting to see some of his mental faculty decline – early-stage Alzheimer's had reared its sinister head. The choice would eventually be a nursing home or our home.

The idea of multigenerational living had intrigued me for years. My *how-did-they-do-it-before* approach to winemaking had grown over the years to cover many aspects of life. Before we put old folks into homes, how did we do it? Of course, in many countries around the world there is no option to put people in homes, and families continue to take care of the aged. It seems sometimes as though we have banished ageing, disease and death to hospitals and care homes. We have pushed it away. In doing so, I wonder if we have diminished our experience of life through family and community, opting instead for the illusion of the individual that lives outside the wild, natural world.

Believe me, I was extremely hesitant about having my in-laws in such close proximity. Like most of us, I had been brought up in that other world, where extended family was more problematic than anything else. Where we each learned to be the centre of our own universes, making it almost impossible to live in close communities, let alone extended families. Where give and take, compromise and patience, and a degree of self-effacement was necessary if the tribe was to function cohesively. Living as a group or even as a basic nuclear family is becoming increasingly difficult in an age when everyone believes they should be an Instagram star. But what if we are missing some fundamental richness of life by doing this? What if my ten-year-old daughter had her aged grandparents living next door? What lessons might she learn about life? What happiness might she find in things as simple as going to collect the eggs from the chicken coop with her grandmother, or walking to collect firewood with her grandfather? What memories might she create, wandering in the vines with them, tasting the grapes in the late summer sun? And what about when things turned difficult and even tragic, as they most certainly would? Should we be shielding children from age, sickness and death? Would she live life more urgently, more fully with that awareness? Would I?

Juliette and I couldn't answer these questions, but we had our opinions. And so, one fine autumn day after harvest, we moved Albert and Bernadette in, back to the farming life that they had lived and left all those years ago.

With the winery clean and closed, autumn's last days unfurl. The leaves on the vines turn their brilliant hues of yellow and fall to the ground, where they will return nourishment to the soil. The days grow shorter, the gates of winter unlock.

I spend time in the cellar, chalking the barrels to identify their contents by parcel and date of harvest. I write the density of the juice in a small leather-bound notebook. If there are empty barrels, we arrange them to be stored for the season and burn the first pastille of sulphur in them, preparing for the long wait until the next harvest when, hopefully, they will once again house wine. I put my ear to the barrels daily, listening for the first signs of fermentation, the music of angels.

Autumn comes to an end for me when, one day without fail, I bend over a barrel and hear the soft fizzing that tells me the yeasts have found their way and are beginning their work of transforming this precious juice into even more precious wine. We did it. Another vintage is on its way, and I look forward with some impatience to following it as it reveals itself, little by little, over the course of the coming seasons.

And of course, the seasons will come. No matter what we do. Year in, year out, they will trace their memories. The rhythm that they beat, their tidal tempo, their changing kaleidoscope of colour and sound, their individual calls to action and inaction, their contrasting revelations of flora and fauna, their music tuned to the cycle of eternal return, and their faithful, symbiotic relationship with these vines and this land. They will give us all of this again and again, asking nothing in return but patience and a modicum of respect.

Acknowledgements

I would like to express my undying gratitude to Vincent and Damien, without whose support and friendship this renaissance, of a vineyard and of a man, would not have been possible.

My gratitude goes also to the many hundreds of people who have come to visit our vineyard over the years and the many thousands of people all over the world who drink our wine. These are the people that inspire us, and feed the passion behind this project. Thank you Rana Dasgupta, for your deep insight and friendship, and Annabel Sherwood for so much inspiration and wise counsel. My appreciation goes also to my UK agent, David Godwin, and to all the good and kind people at Little Toller Books, for their guidance, trust and patience.

Last but not least, my love and thanks to my parents and brothers for their support of this project over the years, both moral and sometimes with muscle, sweat and even a few tears.

Oliver Rackham Library
THE ASH TREE
ANCIENT WOODS OF THE HELFORD RIVER
ANCIENT WOODS OF SOUTH-EAST WALES

Richard Mabey Library
NATURE CURE
THE UNOFFICIAL COUNTRYSIDE
BEECHCOMBINGS
GILBERT WHITE: A BIOGRAPHY

Nature Classics
THROUGH THE WOODS *H. E. Bates*
MEN AND THE FIELDS *Adrian Bell*
ISLAND YEARS, ISLAND FARM *Frank Fraser Darling*
AN ENGLISH FARMHOUSE *Geoffrey Grigson*
THE MAKING OF THE ENGLISH LANDSCAPE *W. G. Hoskins*
A SHEPHERD'S LIFE *W. H. Hudson*
WILD LIFE IN A SOUTHERN COUNTY *Richard Jefferies*
FOUR HEDGES *Clare Leighton*
THE ENGLISH PATH *Kim Taplin*
RING OF BRIGHT WATER *Gavin Maxwell*
COPSFORD *Walter Murray*
THE FAT OF THE LAND *John Seymour*
IN PURSUIT OF SPRING *Edward Thomas*
THE NATURAL HISTORY OF SELBORNE *Gilbert White*

Field Notes and Monographs
AUROCHS AND AUKS *John Burnside*
ORISON FOR A CURLEW *Horatio Clare*
SOMETHING OF HIS ART: WALKING WITH J. S. BACH *Horatio Clare*
THE SCREAMING SKY *Charles Foster*
THE TREE *John Fowles*
NEMESIS, MY FRIEND *Jay Griffiths*
TIME AND PLACE *Alexandra Harris*
EMPERORS, ADMIRALS AND CHIMNEY SWEEPERS *Peter Marren*
DIARY OF A YOUNG NATURALIST *Dara McAnulty*
THE LONG FIELD *Pamela Petro*
SHALIMAR *Davina Quinlivan*
LIMESTONE COUNTRY *Fiona Sampson*
SNOW *Marcus Sedgwick*
WATER AND SKY, RIDGE AND FURROW *Neil Sentance*
BLACK APPLES OF GOWER *Iain Sinclair*
ON SILBURY HILL *Adam Thorpe*
GHOST TOWN: A LIVERPOOL SHADOWPLAY *Jeff Young*
WILD TWIN: DREAM MAPS OF A LOST SOUL AND DRIFTER *Jeff Young*

Anthology and Biography
ARBOREAL: WOODLAND WORDS *Adrian Cooper*
MY HOUSE OF SKY: THE LIFE OF J. A. BAKER *Hetty Saunders*
CORNERSTONES: SUBTERRANEAN WRITING *Mark Smalley*
GOING TO GROUND *Jon Woolcott*
NO MATTER HOW MANY SKIES HAVE FALLEN *Ken Worpole*

Little Toller Books
w. littletoller.co.uk E. books@littletoller.co.uk

North
North Parcel
Barn
West Parcel
Potager
South Parcel
Le Clos de
la Meslerie